INTERNATIONAL REGIMES AND NORWAY'S ENVIRONMENTAL POLICY

International Regimes and Norway's Environmental Policy

Crossfire and Coherence

Edited by
JON BIRGER SKJÆRSETH
The Fridtjof Nansen Institute

Routledge
Taylor & Francis Group
LONDON AND NEW YORK

First published 2004 by Ashgate Publishing

Published 2018 by Routledge
2 Park Square, Milton Park, Abingdon, Oxon OX14 4RN
605 Third Avenue, New York, NY 10017

Routledge is an imprint of the Taylor & Francis Group, an informa business

A Library of Congress record exists under LC control number: 2004008772

ISBN 13: 978-0-815-38981-1 (hbk)
ISBN 13: 978-1-351-15540-3 (ebk)

Contents

Contents

List of Tables and Figures

Tables

Figures

Appendix

List of Contributors

Steinar Andresen is a Senior Research Fellow at the Fridtjof Nansen Institute and Professor at the Department of Political Science, University of Oslo. E-mail: s.e.andresen@stv.uio.no

Hans-Einar Lundli is a Research Fellow at the Department of Sociology and Political Science, Norwegian University of Science and Technology. E-mail: Hans-Einar.Lundli@svt.ntnu.no

Tom Næss is a Research Fellow at the Fridtjof Nansen Institute. E-mail: tom@lnt.no

Marit Reitan is an Associate Professor at the Department of Sociology and Political Science, Norwegian University of Science and Technology. E-mail: maritr@svt.ntnu.no

G. Kristin Rosendal is a Senior Research Fellow at the Fridtjof Nansen Institute. E-mail: kristin.rosendal@fni.no

Jon Birger Skjærseth is a Senior Reaserch Fellow and Research Director at the Fridtjof Nansen Institute. E-mail: jon.b.skjaerseth@fni.no

Jørgen Wettestad is a Senior Research Fellow at the Fridtjof Nansen Institute. E-mail: jorgen.wettestad@fni.no

Preface

This book aims to fill a gap in the literature on international environmental regimes. Instead of exploring the impact of one international regime on its member-states, we look at how one state engages in many regimes to pursue its national goals. As this basic idea was further developed, it soon became clear that governments face a dual challenge in realising their goals on transnational environmental and resource management problems: First, they have to work through international regimes to affect the behaviour of other states and non-state actors; and second, they have to work through national political and administrative institutions to change the behaviour of domestic target groups. This dual challenge implies that most governments face a crossfire of interests and regimes, which demands coherent, coordinated action at both national and international levels.

Our initial idea thus matured into a complex analytical framework. We look at upstream influence directed at 'outsiders' – that is, how Norway is able to work through regimes to influence actors at the international level – by drawing from the study of foreign environmental policy and international cooperation. We look at downstream influence directed at 'nationals' – that is, how the Norwegian government is able to influence domestic target groups – by drawing from the study of domestic political and administrative institutions, with specific emphasis on implementation. And finally, we look at how Norway has been able to link international and domestic efforts over time.

The 'crossfire' project was initiated in 2000 and has developed through a series of workshops at the Fridtjof Nansen Institute. The joint and individual efforts of the research team have been *the* crucial condition for the completion of the project. There is particularly one person to which we all are indebted: Olav Schram Stokke has participated actively in most of the workshops and provided very useful comments to earlier drafts. Many participants in the project team have also benefited greatly from their participation in a related project on interaction between international regimes and EU directives. This EU funded project has been led by Sebastian Oberthür and Thomas Gehring.

Funding for this project has mainly been granted by the Norwegian Research Council to a strategic institute project on international regimes. Crucial additional support was provided by the Department of Political Science, University of Oslo, Department of Sociology and Political Science, Norwegian University of Science and Technology and the Fridtjof Nansen Institute.

Lynn P. Nygaard has helped us with language editing and logical inconsistencies in the manuscript. A special thanks goes to Maryanne Rygg, who standardized references and produced camera-ready copy for the book.

Lysaker, 23 December 2003
Jon Birger Skjærseth

List of Abbreviations

BAT	Best Available Technology
CBD	Convention on Biodiversity
CDM	Clean Development Mechanism
CFC	chlorofluorocarbon
CITES	Convention on International Trade in Endangered Species of Wild Fauna and Flora
CLRTAP	Convention on Long-Range Transboundary Air Pollution
CMS	Convention on the Conservation of Migratory Species of Wild Animals
CO_2	carbon dioxide
CoP	Conference of the Parties
ECE	(UN) Economic Commission for Europe
EEA	European Economic Area
EQO	Environmental Quality Objective
FAO	UN Food and Agriculture Organization
FNI	The Fridtjof Nansen Institute
GATT	General Agreements on Tariffs and Trade
GEF	Global Environmental Facility
GHGs	greenhouse gases
GWP	Global Warming Potential
HCFC	hydrochlorofluorocarbon
ICES	International Council for the Exploration of the Sea
IFF	Intergovernmental Forum on Forests
IMO	International Maritime Organization
IMR	Institute of Marine Research (Bergen, Norway)
INSC	International North Sea Conference
IPC	Integrated Pollution Control
IPPC	Integrated Pollution Prevention Control
IUCN	World Conservation Union
IWC	International Whaling Commission
KBF	Norwegian association of CFC users (*KFK-Brukernes Fellesutvalg*)
ME	Ministry of the Environment
MFA	Ministry of Foreign Affairs
MILJOSOK	Contact forum established by the Ministry of Energy in 1995 involving ministries, oil companies, the fishing industry, research institutes, and environmental NGOs
MoP	Meetings of the Parties
NAMMCO	North Atlantic Marine Mammal Commission
NH_4	ammonia

NINA	Norwegian Institute of Nature Research (*Norsk institutt for naturforskning*)
NMVOC	non-methane volatile organic compound
NO_x	nitrogen oxide
NSCN	Norwegian Society for Conservation of Nature
OAGN	Office of the Audit General of Norway
ODS	ozone-depleting substances
OECD	Organization for Economic Co-operation and Development
OSPAR	Convention for the Protection of the Marine Environment of the North East Atlantic of 1992
PCA	Norwegian Pollution Control Authority (*Statens forurensningstilsyn*), and also Pollution Control Act of 1981
PIC	Prior Informed Consent
PM_{10}	particulate matter
POPs	Persistent Organic Pollutants
SABIMA	Samarbeidsrådet for biologisk mangfold
SBSTA	Subsidiary Body on Scientific and Technological Advice
SIMEN	Report developed by Statistics Norway in 1989
Skogforsk	Norwegian Institute for Forest Research (*Norsk institutt for skogforskning*)
SO_2	sulphur dioxide
TOMA	Tropospheric Ozone Management
TRIPS	(WTO) Agreement on Trade-Related Aspects of Intellectual Property Rights
UES	Uniform Emission Standards
UNCED	United Nations Conference on Environment and Development
UNECE	United Nations Economic Commission for Europe
UNFCCC	United Nations Framework Convention on Climate Change
UNFF	UN Forum on Forests
VOC	volatile organic compound
WCED	World Commission on Environment and Development
WFD	Water Framework Directive
WPA	Water Pollution Act of 1970
WTO	World Trade Organization

Chapter 1

Introduction

Jon Birger Skjærseth

The conventional approach to studying regime effectiveness is to look at how international regimes influence member states in their efforts to make and implement relevant decisions. Regime effectiveness is portrayed as a process of *engaging countries* in the objectives of the regime, whether this objective is the protection of the ozone layer, sustainable management of biodiversity, preservation of whales, or the prevention of hazardous releases into the marine environment.[1] National case studies, which have been the major means of gaining insights in this field, typically use regime objectives as a point of departure. Effectiveness has been measured by the extent to which regime objectives are met (or at least approached) and by the impacts that international rules have on relevant behaviour by regulatory agencies and, ultimately, target groups.

This book takes the opposite point of departure: instead of looking at regime objectives in the context of many member states, we look at national objectives in the context of many regimes. Our perspective is *how states engage international regimes* to pursue national goals within a particular issue area. Sometimes, as in the case of acid rain, there is a high degree of overlap between regime objectives and national goals: Norway's 'pusher' role within this regime stems naturally from the fact that the most relevant target groups are found outside of Norway. If the acidity of Norwegian lakes and forests is to be reduced, the behaviour of target groups in countries that are parties to the relevant protocols under the 1979 Convention on Long-Range Transboundary Air Pollution (CLRTAP) must be changed. At the other extreme, Norway's goals in relation to the International Whaling Commission (IWC) are clearly at odds with the objective of the international whaling regime: whereas the goal of the IWC for the past two decades has been preservation of whales, the key effectiveness criterion for Norwegian foreign whaling policy is whether this policy succeeds in removing (or at least reducing) the legal, political, and economic barriers to small-scale harvesting of minke whales in the North-East Atlantic. Somewhere in between we find Norway's engagement in international regimes where regime objectives figure prominently among Norwegian goals as well, but other and sometimes competing concerns also weigh heavily, such as the need to protect national industries from regulations that would put them at a competitive disadvantage.

How a state engages international regimes to pursue national goals depends first on how domestic political and administrative institutions operate. Who makes decisions about foreign- and national policy, and how are the decisions made and implemented? On the one hand, national authorities face a challenge when it comes to the task of developing national positions in an increasingly complex web of international environmental negotiations. This challenge requires a coherent and well coordinated foreign environmental policy across various issue areas. On the other hand, the same authorities face pressure from domestic target groups, environmental organisations and other interested parties linked to a wide range of domestic sectors, since most environmental problems arise as by-products of otherwise legitimate domestic activities, like production of energy, goods and food. This challenge requires effective domestic implementation of environmental policies. Thus our first main research question is, How does domestic institutionalisation of foreign and domestic environmental policy affect Norway's ability to pursue its environmental goals within specific issue areas?

How one specific state engages an international regime to pursue national goals depends not only on the efforts of that state, but also the efforts of other parties to the regime and the receptivity and strength of the regime. In other words, the endogenous characteristics of the regime itself determine to a certain extent how much influence an individual state may exercise. These characteristics can include the decision rules that apply, and how compliance is enforced. Our second research question is thus: To what extent and how is Norway's ability to pursue national goals within a given issue area affected by the characteristics of the 'core regime'? The term core regime distinguishes international institutions established to deal with specific problems within an issue area from linked regimes (see below). For example, Chapter 4 of this book looks at the Vienna Convention and the Montreal Protocol as the core regime governing the ozone layer. In addition to the ozone regime, we focus on international core regimes established to deal with air pollution, marine pollution, climate change, biodiversity and whaling.

International regimes do not exist in isolation from each other. Linkages and interaction between international regimes is beginning to attract political and academic attention (see e.g. Young, 1996 and Gehring and Oberthür, forthcoming). The international society is populated by more than 200 major international environmental regimes, and this number is growing steadily (Beisheim et al., 1999). Norway is party to over 70 major international environmental agreements that may have a bearing on Norway's environmental policy (Ministry of the Environment, 2002).[2] It is thus reasonable to assume that the manner by which Norway engages an international regime to pursue national goals will depend on interaction between the regime and other linked international institutions.

Linkages between international institutions refers to a situation where individual regimes interact so that other regimes affect the functional scope of the core regime. Such linkages exist either horizontally or vertically (Gehring and Oberthür, forthcoming). Horizontal interaction refers to linkages between 'traditional' regimes, such as between the Convention on Biodiversity (CBD) and the climate regime. Forests are important for biodiversity concerns and they serve as 'sinks' for carbon dioxide (CO_2) (see Chapter 8). In Europe, international institutions are also linked vertically between international regimes and 'supranational' EU legislation. Such vertical linkage may have significant consequences for an individual state. For example, in 1994, Norway joined the European

Economic Area Agreement (EEA), which means that about 80-90 per cent of EU environmental legislation now applies to Norway – even though Norway is not an EU member. Vertical linkage is also visible through international law. In the first half of the 1980s, only 1 per cent of the changes made in Norway's legislation were motivated by international law. In the latter part of the 1990s, this share had increased to 15 per cent.[3] According to Weiss (1993), even industrial states with well-developed environmental regulatory mechanisms and bureaucracies show signs of being overwhelmed.

The consequences of these linkages, either horizontal or vertical, may be either positive or negative – depending on whether the linked institutions work together in line with national goals or contradict each other. Positive consequences occur from mutually reinforcing international commitments where more than one regime requires realisation of national goals in the same issue area. Some issue areas may even require contributions from different types of international environmental institutions and regimes to make international cooperation effective. For example, marine pollution in the North-East Atlantic is simultaneously regulated by international legal conventions, political declarations based on 'soft law', and supranational regulation in the form of EU directives (see Chapter 6). We should also note that divergence in regime objectives is not necessarily a disadvantage for individual states engaged in different regime processes. High regime density can provide good opportunities for 'venue shopping' whereby states can take advantage of linked regimes to promote their own agendas. The Convention on International Trade in Endangered Species of Wild Fauna and Flora (CITES) has, for example, lined up with Norway's goals in the IWC by supporting the Norwegian proposal to down-list the North-East Atlantic minke whale from the threatened species list (see Chapter 3).

Conversely, negative links may stem from consequences flowing from contradictions between different international commitments. Divergence between regime objectives can expose national decision-makers to coordination problems, to such an extent that pursuing national goals in one regime may lead to lower goal attainment in another. For example, some substitutes for ozone-depleting substances are potent greenhouse gases. Ozone-depleting substances may thus be reduced at the expense of increasing emissions of greenhouse gases (GHGs) (See Chapter 4). Thus the third research question that will be addressed throughout this book is: To what extent and how have linked regimes affected Norway's ability to pursue its national goals within particular issue areas?

The discussion above, and our choice of research questions, implies that the crossfire of interests and regimes in developing and implementing environmental policy will require effective coordination of institutions at national and international levels. The focus on institutions is intentional, because we want to look at factors that are possible to manipulate by decision-makers. However, we should stress at the outset that institutions are not necessarily the most important determinant for the extent to which Norway succeeds or fails in pursuing its environmental goals. A state's success or failure in attaining its environmental goals through engaging international regimes may depend on the configuration of interests of other parties to relevant regimes as well as the distribution of costs and benefits among domestic target groups. The issue area itself may be so politically malign in terms of conflicting interests that it precludes effective international cooperation or domestic implementation. Thus we supplement the institutional approach of our case studies with an approach that looks at how problem types might also affect goal attainment.

A Snapshot of Norway in International Environmental Cooperation[4]

The main gist of Norway's approach to international environmental negotiations and commitments can be summed up as follows: Norway has since the 1970s endeavoured to assume the role of a *pusher* state in international environmental negotiations. At the same time, the aim has been to conduct a *pragmatic* environmental policy at the domestic level that would benefit its economic interests. Because Norway faces high abatement costs and contributes little to transnational environmental problems, these two considerations often come into conflict. Though Norway discharges only three per cent of the nutrient flowing into the North Sea, emits a tiny 0.2 per cent of world chlorofluorocarbons (CFCs) and CO_2, and is home to under 0.2 per cent of the world population, various OECD evaluations conclude that Norway's international role in environmental policy is formidable (OECD 1994, OECD 2001).

Norway's green international profile is, however, put to the test when international environmental agreements enter the implementation phase. Indeed, Norway stands out as a paradox for anyone interested in environmental politics. On the one hand, Norway's high profile in international environmental cooperation is closely linked to the magic formula of 'sustainable development' developed by the World Commission on Environment and Development (WCED), led by the former Norwegian Prime Minister Gro Harlem Brundtland. On the other hand, and stated in its extreme form, Norway resembles a huge oil company that is integrated into the business of whale hunting. Norway is highly dependent on oil and gas production and has fought the entire world in order to hunt whales and export whale products.

For oil- and gas-rich nations such as Norway, regulating petroleum tends to maximise the conflict between the need for economic growth and the desire to maintain an environment-friendly reputation. The launching of the report of the WCED in 1987 and the start of a process of formulating a strategy for Norway's leadership ambitions in international environmental politics coincided with the start of a steady and strong increase in Norwegian oil and gas production. Indeed, oil production doubled between the early 1990s and 2001.[5] At this time, the petroleum sector accounted for nearly half of the value of Norwegian exports. This development was accompanied by a steady growth of offshore atmospheric emissions both in terms of CO_2 and air pollutants such as nitrogen oxides (NO_x) and volatile organic compounds (VOCs). For Norway, these growing emissions have highlighted the dilemma of being both a big petroleum producer and at the same time holding ambitions of being a green frontrunner – not least in the field of climate change and international air pollution politics.

Main Policy Lines and Principles

Prior to 1980, global environmental cooperation was weakly institutionalised, and the majority of the regional agreements that did exist were relatively toothless. Traditionally, environmental policy consisted of national and local pollution control and nature conservation. The basic change in international environmental policy in the decades following the 1972 Stockholm Conference coincided with the visible consequences of transnational environmental problems. Forest damage in Germany and fish mortality in Scandinavia in the early 1980s, the discovery of the 'ozone hole' over the Antarctic in 1985, Chernobyl in 1986, the algae 'invasion' in the North Sea in 1988-89, and the Exxon Valdez accident in 1989 were made known through the media and influenced public opinion, particularly in the OECD region.

The WCED was established as a result of UN General Assembly resolution 38/161 in 1983, and the events mentioned above provided the necessary weight and impetus to enable the Commission to make headway. The report *Our Common Future* was the subject of debate in the UN General Assembly in October 1987, and the Assembly adopted a comprehensive follow-up process which led up to the UN Conference on Environment and Development held in Rio de Janeiro in June 1992. If a single event should be recalled from that process, it must be introduction of the notion of sustainable development, which rendered an active environmental policy acceptable to the world at large.

In the shadow of the global process, regional cooperation also altered character, becoming more ambitious over time. Norway and the other Nordic states had actively participated in international cooperation on reducing ocean pollution since the early 1970s. Norway was also an active party in negotiating the Convention on Long-Range Transboundary Air Pollution of 1979. These agreements were strengthened by decisions for significant reductions in emissions in the course of the 1980s. At the North Sea Conference in London in 1987 for instance, a declaration was adopted to reduce discharges of nutrients in sensitive areas and hazardous substances by 50 per cent between 1985 and 1995. Norway has thus not only been obliged to adjust to new global and regional challenges, but also been a prominent participant in this process.

Norway's official foreign policy objectives in the area of the environment became stronger and more explicit as a result of the work of the Brundtland Commission. To quote the report *Our Common Future,* 'The governments not already having done so, would be advised to consider developing a foreign policy for the environment' (World Commission on Environment and Development, 1987, p.226). As a direct consequence of this statement, the Ministry of Foreign Affairs published *Environment and Development. Norway's Contribution to the International Efforts for Sustainable Development* in 1988 (Ministry of Foreign Affairs, 1988). With that, cooperation on 'sustainable development' became the top priority of environmental policy. Furthermore, it was emphasized that this principle would be integrated into all relevant aspects of Norway's international policy. The report also emphasized that Norway would step up, strengthen and improve processes already commenced. At the same time, White Paper 46 (1988-

89) on environment and development, drawn up by the Ministry of the Environment, was published (Ministry of the Environment, 1989). This document also has sustainable development as its point of departure, though emphasis here was placed on Norway being in a favourable position to play the part of driving force in many contexts (ibid., pp.42-43). Most of the goals and principles set out in this first comprehensive Norwegian environmental policy plan still undergird Norwegian environmental policy today.

There has been broad consensus across party lines in Norway to adopt sustainable development as a general foreign policy principle in the area of the environment. Though environmental policy is a relatively new foreign policy area for the Norwegian parliament, it cannot be claimed that this policy departs in any significant degree from the consensual norm in Norway's general foreign policy exercised after World War II (Holst and Heradstveit, 1985). Principles for protecting Norway's own economic interests, promoting binding international cooperation and solidarity through Norwegian aid policy are also familiar from earlier times.

There is nevertheless reason to claim that Norway's international environmental policy contains one important element that differs from most other issue areas: Norway's role as 'pusher'. The term 'driving force' or 'pusher' gives associations in the direction of a strong ideological element and active international leadership. As a small country wedged between the great powers of the world, Norwegians have accustomed themselves to being characterised as 'bridgebuilders' and 'mediators', which also have active cores, though they place Norway in an intermediate position. Though the basis for this role as driving force is not made explicit, it seems to encompass both economic and ideological aspects. First, in terms of the economic aspect, Norway is a 'downstream' country in the regional cooperation, in the sense that Norway is a net importer of both sea-borne and air-borne pollution. As much as 80-90 per cent of the sulphurous and nitrogenous precipitation that falls on Norwegian soil originates in the UK, Russia and central Europe. This means that it is often in Norway's direct economic interest to push for stringent joint international commitments. Second, Norway is negatively affected by its own emissions in some respects. And third, Norway as a country is ranked as one of the richest in the world. The ideological element may best be substantiated through the Government's support of the WCED's notion about the injustice of the overexploitation of resources in relation to coming generations (Ministry of the Environment, 1989, pp.15-16). The ideological element can also be seen in Norway's goal of redressing the imbalance between North and South. (Ibid., p.42).

There are five main principles that constitute Norwegian policy regarding international environmental cooperation:

- First, Norway should lead by example through promoting sustainable development at the domestic level as well as internationally. This involves *environmental concerns being integrated into every sector* that can conceivably contribute to the environmental problems.
- Second, environmental policy towards the world at large as well as domestically should be founded upon *cost-efficiency*. All measures shall be

based upon the principle of achieving a maximum of 'environmental quality' for the money invested.

- Third, Norway shall contribute to making international environmental cooperation more effective. Respect for international law and *adherence to international agreements* are important objectives in this respect.
- Fourth, educated choices shall be made in the sense that political decisions should be based upon *state-of-the-art scientific knowledge*. However, the precautionary principle should apply, meaning that uncertainty should not justify failure to act.
- Fifth, Norway shall *promote development in the Third World*. This aim is closely related to Norwegian aid policy, and in our context will only be mentioned in relation to global environmental problems.

Outline of This Book

The next chapter presents the analytical framework that will guide the empirical analysis. Chapters 3-8 present the case studies. In Chapter 3, Steinar Andresen tells the story of Norwegian whaling. The crux of the whaling case is Norwegian efforts to realise national goals by influencing 'outsiders', since anti-whaling actors have almost exclusively been located abroad. Andresen analyses why Norway decided to halt commercial whaling in the mid-1980s, why whaling was resumed in the beginning of the 1990s, and why export of whaling products was resumed in 2001. The whaling case would remain a mystery to anyone applying a cost-benefit analysis to understand why Norway resumed whaling.

In Chapter 4, Tom Næss presents the success story of Norwegian efforts to protect the ozone layer. The ozone case displays a rare and positive interplay between various explanatory factors: The problem was relatively easy, the core regime was unusually strong, linked regimes were constructive and the implementation strategy and policy instruments applied were adequate.

In Chapter 5, Jørgen Wettestad analyses Norwegian air pollution policies. In this case, Norway has achieved significantly more abroad than at home in terms of goal attainment. As a net importer of air pollution, Norway has, together with other 'pusher' states, worked successfully for stringent international commitments. At home, Norway has faced significant challenges in its efforts to reduce controlled substances in line with international and national goals.

Marine pollution in the North Sea and the wider North-East Atlantic is analysed by Jon Birger Skjærseth in Chapter 6. As in the air pollution case, Norway is a net importer of marine pollution and has worked successfully for a stringent regime that has reduced emissions of regulated substances significantly throughout Europe. Of particular interest in this case is that Norway has become 'trapped' in its own international ambitions as the regime has continued to tighten up its commitments.

The tension between Norwegian achievements abroad versus at home is also pertinent to the case of climate policy, which is presented by Hans-Einar Lundli and Marit Reitan in Chapter 7. Climate change is a global problem where

Norwegian emissions are insignificant. The domestic stabilisation target adopted in 1989 triggered fierce disagreement between various governmental branches on the relative share of GHG emissions to be reduced at home or abroad. With the rise of emissions trading as the main policy instrument, domestic resistance has faded.

Norwegian biodiversity policy is analysed by Kristin Rosendal in Chapter 8. In this case, Norway and the Nordic countries acted as bridge builders in the negotiations between North and South, by stressing that the South should get compensation for the North's use of genetic resources in the South. Domestically, however, conservation of forests lag far behind stated goals, scientific recommendations and other comparable countries.

Chapter 9 presents the comparative analysis and conclusions. It addresses the research questions and the approaches discussed in Chapter 2 in light of the findings from the individual case studies.

Notes

1. For a particularly clear instance, see Weiss and Jacobson (1998).
2. In 2002, Norway was part to 77 major international environmental agreements of which 34 were bilateral agreements. The number rises to 125 agreements if protocols and declarations are included.
3. Notice that most of these revisions did not specifically apply to environmental legislation. See Statskonsult (2000).
4. This section is based on Skjærseth and Rosendal (1995).
5. Oil production increased from a level of around 130 million standard cubic meter oil equivalent (scm oe) in the early 1990s to a level of around 280 scm oe in 2001.

References

Beisheim, M., Dreher, S., Walter, G., Zangl, B. and Zürn, M. (1999), Im Zeitalter der Globalisierung? Thesen und Daten zur gesellschaftlichen und politischen Denationalisierung, Baden-Baden.

Gehring, T. and Oberthür, S. (forthcoming), *Institutional Interaction – How to Prevent Conflicts and Enhance Synergies Between International and European Environmental Institutions*, Ecologic, Berlin.

Holst J.J. and Heraldstveit D. (eds) (1985), *Norsk Utenrikspolitikk*, Tano, Oslo.

Ministry of the Environment (1989), *Environment and Development. Programme for Norway's Follow-up of the Report of the World Commission on Environment and Development* (Miljø og Utvikling. Norges oppfølging av Verdenskommisjonens rapport), Report No. 46 to the Storting (1988-89), Ministry of the Environment, Oslo.

Ministry of the Environment (2002), *Survey of the Most Important International Environmental Protection Agreements* (Oversikt over de viktigste internasjonale miljøvernavtaler), Department for international cooperation, Oslo.

Ministry of Foreign Affairs (1988), *Environment and Development. Norway's Contribution to International Efforts for Sustainable Development* (Miljø og utvikling. Norges bidrag til det internasjonale arbeid for en bærekraftig utvikling), Ministry of Foreign Affairs, Oslo.

OECD (1994), *Environmental Performance Reiews: Norway*, OECD, Paris.

OECD (2001), *Environmental Performance Reiews: Norway*, OECD, Paris.

Skjærseth, J.B. and Rosendal, G.K. (1995), 'Norwegian Environmental Foreign Policy' (Norges miljø-utenrikspolitikk), in T.L. Knutsen, G.M. Sørbø and S. Gjerdåker (eds), *Norwegian Foreign Policy* (Norges utenrikspolitikk), Cappelen, Oslo, pp. 161-80.

Statskonsult (2000), Folkerettens innflytelse på norsk lovgivning. En kartlegging og analyse av gjennomføringen av folkerettslige forpliktelser i Norge, Report 3, Statskonsult, Oslo.

Weiss, E.B. (1993), 'International Environmental Issues and the Emergence of a New World Order', *Georgetown Law Journal*, vol. 81, no. 3, pp. 675-710.

Weiss, E.B. and Jacobsen, H.K. (1998), *Engaging Countries: Strengthening Compliance with International Environmental Accords*, The MIT Press, Cambridge.

World Commission on Environment and Development (Verdenskommisjonen for miljø og utvikling) (1987), *Our Common Future* (Vår felles framtid), Tiden Norsk Forlag, Oslo.

Young, O.R. (1996), 'Institutional Linkages in International Society: Polar Perspectives', *Global Governance*, vol. 2, no. 1, 1-24.

Chapter 2

Analytical Framework

Jon Birger Skjærseth

How does a single country go about attaining its environmental goals in issue areas covered by a web of various international regimes? What are the specific institutional conditions that facilitate – or obstruct – that country's success? Why might the same country have better success meeting its goals in one issue area compared to another? The point of departure for this book is Norway's success (or lack thereof) in attaining its environmental goals in various issue areas both internationally and nationally. Our analysis is guided primarily by an institutional approach that comprises three explanatory perspectives (three groups of independent variables) on goal attainment: domestic institutions, international core regimes, and other international regimes functionally linked to the core regime. The institutional approach is then supplemented with insights gained from an alternative approach that seeks explanation in problem types. Our assumption is that changes and variations in the independent variables identified in both approaches will explain the varying degrees of goal attainment in the different issue areas.

Whenever the actors causing the environmental problems in the first place are located both within and outside a specific country, goal attainment becomes a matter of influencing both foreign and national target groups. The first explanatory perspective reflects this dual challenge. Domestic political and administrative institutions not only play a role in implementing domestic measures, but also help shape foreign policy directed towards international regimes. This perspective draws on contributions from the study of national and comparative environmental policy, and it assumes that administrative arrangements within one state will differ between (or change within) issue areas.

The second explanatory perspective is rooted in well-established approaches to the study of international environmental regime effectiveness. Here, the basic assumption is that goal attainment at the international level depends both on how states exercise influence within core regimes and on the receptivity and strength of such regimes. But international core regimes can also affect goal attainment at the national level in the implementation phase of the regime. Even states that initially push for stringent joint commitments can, under specific conditions, find that such commitments affect goal attainment at home.

The third perspective is based on the recent but growing literature on regime linkages, or interaction. The regime-linkage perspective emphasizes that international regimes have over time come to exist in a context of high institutional density in which one regime is affected by other functionally linked regimes. While previous studies of regime linkages have tended to emphasize problems caused by regime 'congestion', we assume that high institutional density does not necessarily result in duplication of work and coordination problems for national authorities. Regime density can create opportunities for 'venue shopping' in which actors can choose institutional arenas and take advantage of linked regimes to promote their own goals.

Together, these three perspectives form what we call an 'institutional approach'. A second approach, which complements the institutional approach, is based on the literature on problem types, in which issue-area properties and strategic 'games' are central components. In some issue areas, goal attainment can be easier than in others simply because the problem to be solved is more 'benign' in political terms. This perspective focuses on constellations of actor interests and distribution of costs and benefits rather than institutional qualities and reflects that institutions are not the only factor that matters. The main focus of analysis within this perspective is the political malignancy faced by one government in terms of opposing or supporting actors within relevant international regimes and at the domestic scene in different issue areas.

The section below discusses how to define, measure and compare goal attainment in different issue areas.

Focus for Explanation: Norwegian Goal Attainment

The chapters in this book look specifically at the extent to which Norway has been able to attain its environmental goals by participating in international regimes. Thus the focus for explanation, i.e. the dependent variable, is *Norway's goal attainment in specific issue areas.* We have chosen cases in which Norway depends upon the interests and positions of other states and non-state actors in order to realise its goals. Accordingly, the link between foreign and domestic policy becomes important for Norwegian goal attainment. The concept of *target groups* is normally related to non-state actors at the domestic level that need to modify their behaviour in order for a country to realise its national goals. Unless otherwise specified, the concept of *target groups* will in this study refer both to non-state actors and states, i.e. 'target states' that need to modify their policies and behaviour.

The need for international cooperation varies between issue areas, however, according to the degree to which target groups are located inside or outside Norway. If target groups are mainly located outside the country, then goal attainment becomes a question of influencing *outsiders* through international regimes. If target groups are mainly located inside, goal attainment depends on success in influencing *nationals* through the process of domestic implementation.

The criterion applied here for *what can be achieved* is thus attainment of national environmental goals internationally and domestically. Any study of goal attainment within an issue area must be based on a well-substantiated assessment of such goals.[1] Generally, national goals in connection with participation in an international regime tend to be more narrow than those of the regime itself; and the magnitude of this difference is likely to vary with the position of the nation in question in the behavioural system addressed by the regime. For example, the national goals of net exporters of pollution are presumably shaped by conflicting interests to a greater degree than is the case for net importers, and are thus less likely to be compatible with the corresponding regime objectives.

Goals are operationalised interests, and national interests can be either material or ideal, or both. For example, protecting technologically backward cornerstone industries from international BAT (Best Available Technology) requirements is often an implicit goal with a basis in (usually local) material interests. The goal of establishing an international reputation as a 'green champion' in global environmental diplomacy – in this case, the goal of Norway acting as a green pusher – is primarily based on ideal interests. The motive for individual political leaders to assume that role is of course not necessarily idealistic; it can simply result from a positive assessment of how domestic voters will respond, but that assessment will not hold unless a sufficiently large part of the constituency assigns priority to this idealism. Usually, national goals are articulated in terms that combine material interests with ideal interests. Take whaling, for instance. Norway has argued partly in *material* terms, emphasizing that certain small coastal communities depend on the income from whaling activity and citing the economically compelling fact that more North-East Atlantic cod is taken by whales than by humans; and partly in *ideal* terms, arguing that management principles of scarce and renewable resources must be based on scientific advice and guided by the right of a small country to make use of accessible resources in a sustainable way.

With any evaluation, the most important analytical requirement is to make the evaluation criteria explicit. Focusing on explicit goals will minimise the subjective element in empirical analysis compared with more diffuse criteria such as 'the spirit' of an agreement or policy – minimise, but not necessarily eliminate, because national goals can be hard to operationalise. In general, Norway's positions in various international regimes tend to be less specific than domestic goals. But the specificity of domestic Norwegian environmental goals also varies greatly within and between issue areas. Whereas Norwegian goals for preserving biodiversity are vague and explicitly linked to national scientific advice (which is not necessarily consensual and independent from political interests), ozone depletion is mitigated by clear percentage targets based on deadlines and baselines. The problem of *vagueness* will be discussed whenever it occurs in the case studies. In addition, international and national goal attainment will be evaluated separately in order to facilitate comparison (see Chapter 9). Separate evaluation also allows for a systematic assessment of whether Norway has performed better abroad than at home – and if so, why. Moreover, states sometimes pursue different goals internationally than they do at home (see below).

In addition to the problem of vagueness, we also have to deal with the analytical problem of *changes in goals* over time. In political systems sensitive to new social demands – like Norway's open and pluralistic party system – environmental policy goals adopted at different points in time are likely to reflect fluctuations in social demands for environmental improvement. Public pressure is itself a contextual factor that has proven important for explaining outcomes of environmental policy (see Jänicke, 1992; 1997). Since the late 1980s there have been – in line with Down's (1972) theory of 'issue-attention cycles' – significant fluctuations in public values and attitudes towards environmental protection in Norway (see Aardal, 1993; Aardal and Valen, 1995). These fluctuations have affected the income and membership of Norway's green organisations, which are important agents for transforming public environmental attitudes into political influence (Jansen and Osland, 1996; Skjærseth, 2000). It is reasonable to assume that if goals are adopted when environmental concerns are 'down' and the implementation process takes place when environmental concerns are on their way 'up', the probabilities for goal attainment are bright. Similarly, the converse might lead to failure in goal attainment, due to reduced ambitions on the part of the government. Thus, in order to compare goal attainment between different issue areas, we need a common starting point in time. Analytically, this will 'control' for the impact of fluctuations in public pressure on goal attainment, given that changes in public environmental attitudes will be equal across different issue areas.[2]

To meet this requirement of a common starting point, we have extracted the goals formulated in the 1988-89 White Paper on Norway's follow-up to the World Commission on Environment and Development (Ministry of the Environment, 1989) (see also Chapter 1). This report – which was the first comprehensive national environmental policy plan in Norway – will be supplemented by other available sources in order to trace changes in goals prior to the late 1980s. In the cases of air and marine pollution, for example, important goals were adopted prior to 1988 that still have significance for implementation. It also makes sense in methodological terms to evaluate goal attainment with a point of departure in national goals formulated about fifteen years ago. Changing the behaviour of a large number of target groups takes many years, and implementation is normally a matter of incremental change.[3]

Even though we have a common starting point, goals may change after this point in time. Visible environmental crises and changes in scientific knowledge may bring about significant changes in goals. This becomes particularly problematic if the level of ambition is reduced within the same issue area. Since lower ambition means greater probability for goal attainment, we may actually face situations where high ambitions and medium goal attainment is judged as a lower degree of achievement than low ambitions and high goal attainment. Let us take Norwegian climate policy as an example: Between 1989 and 1995 Norway aimed at stabilising CO_2 emissions at 1989 levels by the year 2000.[4] In 1995, the climate target was apparently set aside until a new and weaker stabilisation goal (to be achieved by 2005) was re-adopted in 1997. Against which standard should we then evaluate Norway's ability to pursue its climate target at home? In other cases, the goal itself may change from material to ideal interests. Norway's whaling goal has changed – from maximising profits from whaling to the political right of self-determination based on science, sustainability and multi-species ecological management. There are no easy solutions to these analytical problems, but we will apply the following rules of thumb: First, any change in goals or ambition levels will be discussed explicitly in each case study. Second, we will give precedence to international obligations to which Norway has committed itself. If the level of national policy ambitions is reduced 'below' the level of international obligations that Norway has voluntarily approved, we will take this as an indication of 'failure' in goal attainment, or non-compliance.[5]

What has been achieved will be measured first within a set of case studies based on issue areas covered by a core regime (Chapters 3-8), and then (in Chapter 9) between issue areas and core regimes. The prominence of national implementation processes in these case studies will vary with the relative significance of domestic and foreign target groups in the activity system addressed by the regime. In a conventional means–ends hierarchy, the commitments of international regimes are typically a means to achieve other ends, such as removing barriers to harvesting minke whales. In the whaling case, the relevant target groups are primarily non-Norwegian decision-makers and interest groups that oppose whaling. Hence, Norway's interest in the whaling negotiations in the International Whaling Commission (IWC) has centred on foreign target groups. By implication, success in the issue area of whaling will be measured not by changes in Norway's implementation of international commitments towards national fishermen and whalers, since only limited change has occurred in this relationship, but by the increased leverage provided by regime rules regarding the acceptable scope of whale hunting and the export of whale products. This is referred to as goal attainment *internationally*, as measured in terms of the congruence between Norway's foreign policy positions and regime objectives, as well as change in behaviour of foreign target groups wherever feasible. In cases where the primary non-state targets are located outside Norway (as with acid rain), strict international rules are the most relevant means for influencing the behaviour of foreign target groups.

In other issue areas where domestic target group behaviour is highly significant to Norway's goals in the issue area, such as marine pollution, it will be important to assess the national implementation of commitments under the relevant international regimes when measuring national goal attainment. This will be referred to as goal attainment *domestically*, as measured in terms of the congruence between environmental policy goals and the behavioural change of relevant domestic target groups.

Goal attainment will normally depend on a combination of international and domestic efforts, since most environmental problems have an element of interdependence rooted in transnational environmental consequences. This is reflected in the large number of international environmental regimes. Seen from the perspective of any one state, successful environmental policy will thus depend on its ability to affect target-group behaviour both abroad and at home.

Explaining Goal Attainment

International cooperation aimed at promoting a clean environment or sustainable management of natural resources is based on interdependence, since it is only under such conditions that environmental cooperation makes sense. Norway's ability to pursue its environmental goals thus depends upon a combination of its own efforts and the efforts of other actors within relevant regimes.

The explanatory approach presented below is not exhaustive, but reflects our interest in institutional arrangements at the international and national levels. As noted, we will examine three clusters of institutional variables to explain Norwegian goal attainment:

- domestic administrative and political institutions
- international core regimes
- other international regimes functionally linked to core regimes

The emphasis placed here on institutions at the international and national levels implies that in order to understand variation in goal attainment, we have to combine the two institutional strands of political science: the one focusing on domestic political and administrative institutions, and the other dealing with international relations (regime theory) (Powell and DiMaggio, 1991, p.5). The term 'institution' has varying meanings within different social science disciplines (Hall and Taylor, 1994).[6] In this study, it is used to refer to constellations of rights and rules that define social practices, assign roles to participants in those activities, and guide interactions among those who occupy those roles (Young, 1994, p.3). Rules constitute the substantive component of institutions and may operate at two levels. At the 'micro' level, institutional rules should be understood as prescriptive statements that forbid, permit or require some action or outcome (Ostrom, 1990). At the 'macro' level, rules refer to the 'rules of the game' aimed at handling collective choice situations.

Within the broad field of institutional analysis applied to international environmental cooperation and national environmental policy, we will draw primarily upon three strands of literature to understand the impact of domestic institutions, core regimes and linked regimes: (1) the study of national and comparative environmental policy, (2) regime theory, and (3) the evolving literature on interaction or linkages between international regimes. Below, we relate these three clusters of institutional variables to the location of target groups, i.e. whether they are located abroad or at home. The logic behind this analytical distinction is that institutions need to fulfil different *functions* according to the location of target groups. When target groups are located abroad, the relevant question is how institutional factors are likely to affect Norway's ability to pursue its goals internationally by influencing 'outsiders'. At the domestic level, for example, this function will require an effective foreign environmental policy. Conversely, when target groups are located at home, the relevant question becomes how institutional factors are likely to determine Norway's ability to pursue its goals domestically by influencing 'nationals'. This function will at the domestic level require effective integration of environmental concerns into affected sectors, as well as adequate policy instruments.

The institutional perspectives referred to above are largely described in static terms. We know that institutional arrangements at domestic and international levels established in early phases can affect later events (see e.g. Young, 1989; Underdal and Hanf, 2000; Young, 2002). The dynamic element of institutions may lead to path-dependent processes affecting goal attainment at home as well as abroad. In international environmental cooperation, a long period of time may elapse from the development of foreign policy positions to domestic implementation. At the domestic level, decisions taken initially as to who should participate how and when in the development of national positions in core regimes can have important consequences for the prospects of successful implementation at later stages. At the international level, international regimes can develop over time and new regimes can emerge, interrupting initial expectations as to what is needed to achieve stated goals. The dynamic element of institutions will thus be included in the explanatory perspectives.

As noted, institutions are not all that matters, and sometimes they are not even the most important explanation of goal attainment. In addition to institutions, the characteristics of the problems themselves constitute a well-established alternative explanatory approach that can supplement the institutional approach.

Domestic Institutions

National environmental policy can be seen analytically as a function of social demand for environmental quality and governmental supply of policies to protect the environment (Underdal, 1995; Underdal and Hanf, 2000). The analysis in this book looks primarily at the supply side.[7] Governmental supply of policies may go 'upstream' in the form of foreign-policy formulation of national positions in international cooperation, or 'downstream' in the form of implementing international regime commitments by forming and carrying out domestic policy. The upstream process aims at influencing 'outsiders', while the downstream process aims at influencing 'nationals'.

The upstream process: Influencing outsiders. The study of foreign policy processes in the field of environmental cooperation has been a relatively neglected area of environmental policy research. The literature on environmental foreign policy is, however, growing, and it is based on a variety of different perspectives spanning from leadership theory to the assumption of states as unitary rational actors (for a review of this literature, see Barkdull & Harris, 2002).[8] One prominent approach reviewed by Barkdull and Harris views comparative foreign environmental policy as a product of institutional arrangements at the national level.[9] This approach is largely compatible with the Domestic Politics model developed to understand negotiating positions in relation to international environmental cooperation (Underdal and Hanf, 2000).[10] In contrast to the Unitary Rational Actor model, the Domestic Politics model views governments as complex organisations over which no single decision-maker has full control. One distinctive feature of Norway's environmental foreign policy making is for example that the Ministry of

Foreign Affairs plays a varying, often subordinate, role in the formation of national positions in different international regimes (Skjærseth and Rosendal, 1995).

Environmental problems and policies affect a whole range of sectors of society. In the case of climate change, for example, many sectors are affected – from households via transport to agriculture. The basic premise is that different branches of government will tend to perceive environmental problems differently and will therefore apply different decision criteria to such problems – a premise well captured by Allison's (1971, p.176) aphorism, 'where you stand depends on where you sit'.[11] Thus, if you sit in the Ministry of the Environment you will typically put emphasis on environmental damage, whereas those sitting in the Ministry of Finance will tend to focus on (short-term) abatement cost, and the Ministry of Foreign Affairs will emphasize how environmental goals relate to other national goals and interests in other issue areas, such as trade. The point is that none of these preferences will necessarily be fully consistent with environmental goals in specific issue areas.

The distribution of competence and influence between various branches of government and the unity of the cabinet have been found to be important for explaining national environmental policy in relation to international environmental agreements (Underdal and Hanf, 2000). The main assumption to be explored in this study is that Norway's ability to influence 'outsiders' depends on the degree to which different branches of government have been able to coordinate their positions so as to frame a solid and *coherent* foreign environmental policy. A coherent position is important because internal division can be exploited by other states promoting opposing positions. This might weaken Norway's goal attainment internationally. The need for coordination is likely to increase, and the probability of a coherent position is likely to decrease, the more ministries and agencies that are involved in the 'upstream' process. Level of coherence is likely to be determined by, first, the number and types of branches involved, and second, the distribution of competence among those branches. If competence is shared between different ministries responsible for the same regime, coordination problems are likely to arise.

The downstream process: Influencing nationals. The disjunction between the coherency of ecosystems and the fragmentation of political and administrative institutions for implementing environmental policy has often been criticised. The 'cure' suggested is some version of integrated policy, or environmental policy that can penetrate all policy levels and all governmental agencies involved in its execution (Underdal, 1980b; Jänicke, 1992; Weale, 1992, p.94).

In sharp contrast to the fragmentation witnessed between environmental treaties at the international level, domestic environmental policy in OECD countries has been marked by efforts to *integrate* environmental concerns into the sectors that caused the problems in the first place, such as transport, energy and agriculture. This calls for firm coordination of environmental policy. The idea of transferring environmental responsibility and authority to sector agencies can be traced back to the recommendations of the World Commission on Environment and Development (1987, p.19). These recommendations were followed up by

Agenda 21 linked to the 1992 Rio Summit and have subsequently led to follow-up measures in many countries (Weale and Williams, 1993).

There are essentially two routes to sector integration at the ministerial level (Wilkinson, 1997): giving the ministry of the environment authority over other relevant ministries, or providing environmental competence to the sector ministries. The latter route is most common in OECD countries, including Norway.[12] Sector integration is a question of the extent to which relevant environmental goals and follow-up procedures have been coordinated and incorporated into the decision-making processes of affected sector ministries and agencies. Integration of environmental concerns takes place by establishing organisational structures, cross-sectoral laws such as impact assessment, and coordination procedures in the form of inter-ministerial bodies.

Coordination of national policy and sector responsibility can create institutional tension, which represents a challenge to most administrative systems. The Norwegian system mirrors this challenge: on one hand, the principle of sector responsibility enjoys high priority in the country's environmental policy; on the other hand, the environmental authorities are responsible for coordinating national and sector goals through the government. Notice that there is no necessary consistency between those sectors involved in the shaping of foreign policy and those involved in the implementation phase (see below).

Insufficient policy integration between governmental branches can lead to *horizontal disintegration* – whereby the aggregate of decisions in various governmental branches may deviate from stated goals. Against this backdrop, we assume that the more sectors that are affected in one issue area, the more difficult it will be to integrate relevant goals and measures in affected governmental branches. By implication, goal attainment will become difficult. Governments also need to affect societal actors in order to pursue their goals. Since non-state target groups often 'control' the behaviour that has to be modified, governments depend upon their reluctant or active cooperation. A well-known proposition states that the aggregate of 'micro-decisions' among target groups may deviate more or less systematically, or substantially from higher-order policy goals, thus leading to *vertical disintegration* in environmental policy (Underdal, 1979).

Changing the behaviour of domestic target groups requires policy instruments that are adequate with regard to the sources that cause the problem in the first place. Policy instruments can be defined as the means by which public authorities seek to alter the behaviour of target groups; they represent the 'sharp end' of the policy process (Jordan, 1999). This means that vertical disintegration of environmental policy can be counterbalanced by the adoption of adequate policy instruments. Policy instruments have been categorised in several ways – the most familiar distinguishing between regulation, information, and economic instruments (or the stick, the sermon, and the carrot (Vedung, 1997). The government may force us, persuade us, or pay us (or make us pay). In addition, so-called voluntary agreements between the state and target groups have become increasingly popular in environmental policy. Such agreements between the state and target groups are seldom entirely voluntary, since they are negotiated in the 'shadow' of hierarchy. They vary from pure gentleman's agreements to agreements linked to a broader legal context. This in turn means that the extent to which they represent an alternative to regulation will also vary. However, one advantage of voluntary agreements is their capacity to create constructive cooperation between authorities and target groups, cooperation that may influence the norms of environmental behaviour (Skjærseth, 2000).[13]

These types of policy instruments have differing strengths and weaknesses, depending upon the criteria against which they are assessed as well as the characteristics of the problems they are intended to solve. The most relevant criterion applied in this study is the capacity for goal attainment, indicating whether goals are actually achieved at the right time.[14] We assume that the adoption of adequate policy instruments in terms of goal attainment will be more difficult when goal attainment is conditioned by behavioural change in diffuse sources, as opposed to point sources. Point sources – such as industrial facilities – can be effectively controlled by regulation in the form of permits. By contrast, diffuse sources are characterised by a high number of targets, such as car drivers or farmers.[15] When environmental problems are caused by different types of sources, a broad-based portfolio of policy instruments is needed. Goal attainment is likely to become most difficult under such circumstances.

Linking foreign and national policy. The previous sections explored how governmental branches and target groups can affect foreign and domestic policy in separate phases. This section focuses on the linkages between these phases of environmental policy, since institutional arrangements established in an early phase may be important for outcomes at later stages (Young, 1989; Underdal and Hanf, 2000). The basic assumption is that the prospects for successful goal attainment depend on *access* to decision-making for affected branches of government and target groups: to what extent were the agencies and target groups responsible for carrying out implementation involved in formulating the negotiation positions and national goals at an earlier stage?

It is reasonable to assume that early participation by stakeholders will enhance their support for the policy in question and provide valuable information about the consequences of regulation. After all, if affected actors have not had their say, their interests are less likely to be reflected in relevant policy. Since environmental regulation frequently implies net costs for target groups, the strategy of these groups will often be aimed at watering down governmental regulations. In the EU for example, a majority of companies perceive environmental regulations as having the greatest adverse effect on competitiveness after tax and employment regulation (Grant and Newell, 2000).[16] When decisions are to be implemented, resistance and opposition may become severe if regulations are directed against their interests. Conversely, broad access for governmental branches and target group sectors tends to ensure that their interests will be reflected in goals, or the means by which goals are to be achieved. Support for national policy is thus more likely when target groups have had their say. If, for example, environmental goals seriously affect the interests of the agricultural sector, the probability for successful goal attainment – perhaps at the expense of less ambitious goals – will increase if farmers' organisations and ministries and agencies of agriculture have had a say in formulating those goals.

Taking into account the phases of foreign policy formation shows how decision-makers face various dilemmas. First, while broad participation is likely to ensure goal attainment, this inclusiveness in decision-making may also lead to a lower level of ambition in the first place. It is thus an open question whether broad participation will lead to higher environmental effectiveness than narrow participation. Second, broad participation in the formulation of national positions will increase the risk of internal divisions in foreign environmental policy. Finally, decision-makers tend to operate within a 'veil of uncertainty' (Young, 1988). Uncertainty with regard to affected actors, alternatives and consequences may initially facilitate agreement on goals, but the extent to which such uncertainty facilitates implementation will depend on what decision-makers actually see at a later stage – especially with regard to who is affected and in what way.

International 'Core' Regimes

The upstream and downstream processes are related to the development or operation of international regimes. First, foreign policy positions are directed toward international negotiations aimed at establishing, amending, developing or linking international regimes in order to affect 'outsiders'. Most environmental issue areas in which Norway is involved are covered by a core international regime, such as the International Whaling Commission (IWC) or the Convention on Biological Diversity (CBD) and related cooperative arenas. Such regimes tend to emerge or develop from a bargaining process among a group of countries seeking to reach joint international commitments (Young, 1989; Underdal and Hanf, 2000).[17] Second, international commitments produced by core regimes must be implemented domestically to have an effect on 'nationals'. Ultimately, it is decisions at the national and target-group levels that affect target-group behaviour, goal attainment and environmental quality.

The international system has been characterised as anarchy based on a self-help system among states (Waltz, 1979). States can, however, modify the self-help element of international relations by transferring authority to international regimes. The extent to which and how international regimes make a difference has been a contested issue within the grand theoretical debates on international relations as well as in the study of international environmental institutions (see e.g. Grieco, 1988; Mearsheimers, 1995). Sometimes, bargaining never progresses beyond the interests and preferences of the least ambitious actor. Joint international commitments may simply reflect the lowest common denominator of the interests and preferences of the actors, a phenomenon known as the 'law of the least ambitious program' (Underdal, 1980a). The possibility of actors adopting 'dead letters' is compatible with some of the criticism that international regimes merely reflect the distribution of interests and power of the relevant actors (Mearsheimers, 1995). However, a growing body of theoretical and empirical literature has shown that international environmental regimes in different issue areas do indeed have an independent effect on national policy and behaviour (see e.g. Levy et al, 1995;

Miles et al, 2002). Sand (1991) has specified several 'fast track' options that can beat the 'law of the least ambitious program', and Zürn (1991) has distinguished among fifteen groups of variables that can be affected by international institutions.[18] In particular, joint international commitments potentially constrain activities and limit discretion at the domestic level through various behavioural mechanisms, including norms and incentives (see e.g. Ruggie, 1983; Nollkaemper, 1993; Skjærseth, 2000).

It seems logical that international commitments can progress beyond the positions of the least ambitious states – but it is less obvious that joint international commitments can progress beyond the positions of 'pusher' states, such as Norway (see Chapter 1). The most ambitious actors will push for stringent commitments for ideal or material reasons – to improve environmental quality or 'upload' their regulatory standards to the international level to establish equal competitive conditions for national industries. Before turning to the question of whether and how international regimes can progress beyond the goal of even the most ambitious state, let us explore the upstream processes by which Norway can affect 'outsiders'.

Influencing outsiders. It is reasonable to assume that Norway's ability to pursue its goals in core regimes will depend on at least the following three factors:

- Norwegian efforts
- the efforts of other actors
- the receptivity and strength of the regime itself

The relative effort of other actors will be addressed in the section on problem types below. Norway's influence on joint commitments depends first on the level of activity and *means* by which Norway seeks to exercise its influence internationally, and second on the *receptivity* of the regime. The degree to which the regime can affect foreign target groups will depend on the *strength* of the regime. Adequate means of influence, high regime receptivity and strong regime commitments are thus assumed to increase the probability of goal attainment internationally.

With regard to *means of influence*, we first assume that the more active a state is in furthering its interests in the core regime, the higher the probability of its success in 'uploading' its goals to the international level. A high level of activity can sometimes take the form of leadership that will increase the probability of influence. The literature on *leadership* suggests various types of leadership and mechanisms through which influence is exercised (Young, 1991; Underdal, 1991; Malnes, 1995).[19] One important distinction differentiates between 'instrumental' and 'coercive' leadership: the former type is related to negotiating skill, high activity in proposing regulations and the ability to develop integrative solutions, while the latter is related to some notion of power. Normally, negotiating skill and ability will represent a more accessible tool for small states than coercive power, or leadership by unilateral action (i.e. when an actor sets the pace to which other parties may find themselves more or less compelled to adapt) (Underdal, 1991).

There is, however, another type of leadership that may be equally relevant. Intellectual leadership refers to one who produces intellectual capital that may shape or affect the perspectives of others (Young, 1991). The causes and consequences of environmental problems are closely related to the natural sciences, so the linkage between science and policy has become an integral part of most international environmental institutions (Andresen et al., 2000). Intellectual leadership is likely to be exercised when the country in question possesses exceptionally high scientific capabilities and knowledge in specific issue areas, such as marine biology or atmospheric sciences.

A high level of activity cannot be taken for granted. Different negotiating positions can be categorised according to the ambitiousness of preferred regulations in relation to regime objectives. 'Bystanders' connotes a familiar category of states that participate in international environmental cooperation. Bystanders do not relate preferred regulation to regime objectives, but sit on the fence during the negotiations. A low level of activity may be due to various reasons, spanning from limited economic, political or ideal interests in the activity regulated by the regime to lack of scientific or administrative capacity. 'Pushers' advocate negotiating positions that are more ambitious than existing regime objectives, while the 'laggards' advocate positions that are less ambitious. In between, we find the 'intermediates', who advocate positions in line with regime objectives. It should be borne in mind that one and the same state can be a 'pusher' in one regime, a 'laggard' in another and a 'bystander' or 'intermediate' in a third.

Since the *receptivity* of the regime in question affects the extent to which one state will be able to influence joint commitments, the same level of efforts or leadership may be sufficient to succeed in one regime, but insufficient in another. Norway, for example, found itself forced by the IWC to stop whaling at one point in time, whereas it succeeded (together with others) in establishing an international regime to counter acid rain (see Chapters 3 and 5). Receptivity will here be seen as a product of decision rules of the 'core' regime. Decision rules represent a key variable among international regime scholars as well as among decision-makers (see e.g. Underdal, 1995; Wettestad, 1999; Ministry of Environment, 1989). In the 1988-89 White Paper on Norway's follow-up to the World Commission on Environment and Development, Norway adheres to new legal principles that may strengthen international environmental cooperation and make international treaties more effective. One of these principles involves establishing a new international authority on protecting the atmosphere (particularly climate change) that would adopt decisions by majority ruling.

Seen from the point of view of a single state, decision rules should meet at least two contradictory criteria. First, it should be possible for a specific state to veto regime progress whenever proposals from the majority are in conflict with the interests and preferences of that state; and second, it should be possible for a state to outvote the minority whenever its interests and preferences are in line with the majority. The first criterion is met when *unanimity* is required, as decision rules based on unanimity effectively provide each member-state with veto power. Unanimity will, however, lead to the 'law of the least ambitious program' in the absence of issue linkages. The second criterion is met under condition of a *simple or qualified majority*, which provides each member state with the power to outvote the minority. Even though formal voting may be rare in international environmental cooperation, studies of qualified majority voting in the EU indicate that the possibility to do so creates pressure to make concessions – a phenomenon described as the 'shadow of the vote' (Weiler, 1991).

These contradictory criteria cannot be met simultaneously, and since interests of the participating parties will differ, it will be impossible to find a decision rule that will work to the advantage of every state at any given time. However, since the interests of one specific state are likely to differ across various issue areas covered by international regimes, we assume that *consensus* is the decision procedure that will *generally* serve the interests of each member state most of the time. Consensus does not grant veto power (that limits measures to the least enthusiastic party), but it requires the absence of objections. This means that a specific decision is binding for those actors in favour of that decision. Sand (1991) has, as noted, identified 'fast track' options compatible with consensus that can beat the 'law of the least ambitious program' including the promotion of voluntary over-achievement, selective incentives and the principle of differential obligations and regionalisation.[20]

Whereas regime receptivity determines the scope for influence, the *strength* of the regime in question will determine its impact on foreign target groups. The extent to which regime strength can facilitate goal attainment for one specific state will depend on the influence of that state in the regime and consequently the level of congruence between national positions and regime objectives. Regime strength can be seen as a function of *legal status*, *specificity*, *level of ambition* and *compliance control*. *Legal status* centres on whether joint commitments are binding within the framework of international law, where the fundamental norm is *pacta sunt servada* (treaties are to be obeyed). Chayes and Chayes (1993, pp.85-86) argue that such norms exist in international affairs and that a growing body of empirical study and academic analysis supports this view. According to Stokke and Vidas (1996, p.23), the legitimacy of an international regime can be defined as the persuasive force of its norms. Hence, we assume that legally binding commitments – within the framework of international law – may increase the probability of changing the behaviour of foreign target groups simply because these commitments are binding. *Specificity* determines the extent to which the actors know precisely what to do. Young (1979, p.99) refers lack of specificity as the problem of 'operationality'. If obligations are vague, this will in itself contribute to discretion due to interpretation problems; it will make the actions of other states less predictable; and it may also make verification of state behaviour more problematic.

Level of ambition refers to the 'amount' of behavioural change that is required. And finally, *compliance control* is regarded as a function of verification and enforcement. Verification centres on monitoring of actor performance, while enforcement refers to measures aimed at promoting compliance with international commitments. Verification systems to check that other actors are complying with their obligations will make it easier to identify defectors. Credible verification systems increase transparency and are likely to stimulate mutual expectations that cheating will be detected, thereby also removing incentives for free riding. Procedures for dispute settlement may check whether states are staying within the intended scope of discretion. Enforcement systems may provide additional incentives for states to comply with their obligations. Such systems may make desired behaviour less costly when obligations are to be implemented, or they can be more negative, aimed at making undesirable behaviour more costly. Notice that compliance control in a broader sense is often framed in terms of two alternative strategies: enforcement versus management (Tallberg, 2002). While enforcement strategies focus on changing behaviour as described above, management strategies emphasize capacity building, which is most relevant with regard to developing countries.

This book cannot cover in full the causal links between these factors related to regime strength and the behaviour of foreign target groups.[21] What we aim for in the case-studies is to explore the extent to which there is a relationship between the strength of a regime and the way in which foreign target groups behave.

Influencing nationals. It is by no means obvious that a state that initially pushes for stringent joint commitments will in turn be affected by the same commitments. Under such circumstances, the arrow of influence is likely to go from that state to the regime. However, international regimes can influence even the most ambitious actors, under two specific conditions. First, countries can advocate foreign policy positions that are more ambitious than domestic goals. The will to advocate joint and stringent international action does not necessarily imply that a state has a correspondingly strong incentive to comply with its own commitments. As Weale (1992, p.48) has pointed out, governments may be highly reluctant to embark upon policies concerning international environmental obligations. Since transnational pollution problems are frequently asymmetrical in the sense that actors are affected differently by the actions of others, a given commitment may be less favourable than what an actor would have chosen by itself. For example, a net importer of pollution such as Norway would probably prefer that others contribute more towards reducing discharges (notably the exporters), while the actor itself contributes less. Under these circumstances, a strong international regime can 'backfire' and place significant pressure on pusher states. As a response to such regime pressure, states can comply or choose to defect from joint international obligations.[22]

Second, international regimes can develop in stringency over time from the point when initial national positions and goals were adopted. Institutionalised cooperation can, under certain conditions, gather momentum through a 'snowball' effect that generates positive feedback and thus facilitates further steps. According to Young (1989), international regimes can evolve continuously in response to their own inner dynamics. According to this perspective, joint commitments will over time become more ambitious and increasingly place more pressure on governments to implement them. The 'snowball' effect is most likely to develop when the initial institutional arrangements have a narrow scope, include lenient commitments and possess institutional feedback mechanisms that encourage dynamic development (Skjærseth, 2000). On the other hand, and under different conditions, institutionalised cooperation can also stagnate or recess.[23]

This line of reasoning implies that we have to combine two factors in order to understand the impact of core regimes on domestic goal attainment in pusher states. First, under the conditions specified above, a strong core regime – as measured in terms of legal status, specificity, level of ambition and compliance control – can affect domestic target groups (as well as foreign target groups). Second, the extent to which regime strength affects domestic target groups in line with stated goals will depend on the congruence between regime objectives and domestic policy goals. If national goals are going in the same direction as regime objectives, but are less ambitious than international commitments, a strong regime will promote compliance with international obligations and thus increase goal attainment.[24] If national goals are significantly at odds with regime objectives, a strong regime can reduce the level of goal attainment. The development within the IWC from the 1960s until Norway was forced to quit commercial whaling illustrates this possibility (see Chapter 3).

In a situation characterised by strong regimes and high congruence between regime objectives and domestic environmental policy, the core regime itself will have limited influence on implementation and goal attainment. The arrow of influence will actually go the other way: Norway will seek to influence regime commitments by referring to its higher domestic regulatory standards or prior achievements that are more substantial than those agreed internationally. Still, strong regimes may under such circumstances improve the international competitive situation for domestic target groups, such as industry. Strong regimes and low congruence between regime objectives and domestic goals can, as stated above, point to either reduced goal attainment, or compliance with international obligations going in the same direction as national goals. In either case, international regimes are likely to have a significant impact on goal attainment. With a weak regime and high or low congruence, we assume that the regime will have little or no influence on goal attainment. In these cases, the regime is unlikely to influence the international competitive situation for target groups. These situations may be fleeting, since the regime may evolve toward more stringent commitments over time.

International regimes can affect domestic environmental policy and ultimately target-group behaviour in various ways. The relevant literature indicates at least two main (although not mutually exclusive) pathways through which international institutions may have an impact. (see e.g. Cortell and Davis, 1996). First, international rules may become institutionalised into the domestic political process by being incorporated in national law. The *legal pathway* has the capacity to affect not only the level of ambition of domestic policy, but also the relationships between governmental actors, green organisations and target groups. Second, government officials and societal actors can invoke international commitments to further their own specific interests in domestic policy debates. The *political pathway* may thus empower actors in domestic policy debates. Hence, international regimes may serve as 'agents of internal realignments' (Levy et al, 1995, p.307). In essence, international regimes might affect the alignment of domestic groups endeavouring to influence a government's behaviour. Likewise, governments may also be empowered to take action. Government officials may appeal to international rules in order to legitimate unpopular decisions directed towards reluctant target groups.[25] In essence, the existence of international commitments may be utilised for purposes of justifying an actor's own actions, or to question the legitimacy of the actions of other actors.

Linked Regimes

The analysis of international environmental regimes and their effectiveness has traditionally been based on the assumption that international regimes exist in isolation from each other (see e.g. Miles et al, 2002). More recently, regime analysis has been expanded to include the consequences flowing from the linkages between various regimes (see e.g. Weiss, 1993; Young, 1996; King, 1997; Rosendal, 2001; Stokke, 2000; Andersen, 2002; Skjærseth, 2003a; and Gehring and Obertür, forthcoming). This new strand of analysis, which is closely related to the regime effectiveness literature, recognises that most regimes are linked, directly or indirectly, to other international regimes in which the same states often participate. Many efforts to analyse such linkages have tended to emphasize the problems caused by regime 'congestion' (see e.g. Weiss, 1993). Despite the growing scholarly interest in regime linkages, our knowledge remains limited, and scientific exploration is at its early stages.[26]

If the study of how international regimes interact, and with what consequences, can be said to still be in its formative stage, the study of how linked regimes affect environmental policy as seen from the perspective of one country is even less developed. It is reasonable to argue that the ultimate test of the extent to which and how regime linkages produce positive or negative consequences for goal attainment or effectiveness should be undertaken from the bottom up, i.e. from the perspective of single states under crossfire from an increasing number of international agreements.[27] In this section, we shall take one exploratory step in that direction.

Young (1996) distinguishes between four categories of what he terms regime interplay: embedded regimes, nested regimes, clustered regimes and overlapping regimes.[28] We limit our analysis here to overlapping regimes, defined as a situation where individual regimes, formed for more or less different purposes, interact so that other regimes affect the functional scope of the core regime. More specifically, regime linkages can affect the development and performance of the core regime, including its effectiveness. Such linkages can be distinguished according to various policy fields, causes, consequences and policy responses (Gehring and Oberthür, forthcoming).

Seen from the perspective of the individual state, regime linkages can lead to both disruptions and new opportunities. The latter phenomenon is perhaps best understood in terms of 'venue shopping', where actors can choose institutional arenas and take advantage of existing regimes to promote their own agenda (Young, 2002). According to Young, actors can capture, reform or integrate existing institutions. The extent to which regime linkages will enhance goal attainment is determined by a combination of two factors: how (successful) actors exploit such linkages, and the characteristics of the institutional setting in which the actors operate. The remainder of this section explores some likely consequences for goal attainment of congruence or divergence between linked regime objectives and national policy.

First, *divergence* between linked regime objectives can expose national decision-makers to coordination problems, to such an extent that pursuing obligations in one regime may lead to a lower degree of goal attainment in another. It is especially at the interface between trade regimes and environmental agreements that such problems have been in focus. For example, potential conflicts between the international trade system and international environmental treaties have repeatedly been discussed in the WTO Committee on Trade and Environment. However, coordination problems can also occur between environmental regimes. One example is the potential conflict between marine pollution regimes and the climate regime on the issue of CO_2 injection into the seabed. Such injection may reduce CO_2 emissions but violate international obligations on dumping at sea. On the other hand, we should also note that for a state engaging in several international regime processes that are functionally connected, divergence in regime objectives is not necessarily a disadvantage. Indeed, it may even be highly expedient for a state to differentiate its policies in different regime processes. Whenever there is a mismatch between the objectives of the core regime and the goals of one specific state, that state has the possibility of furthering its interests in functionally linked regimes. High regime density and good opportunities for 'venue shopping' can thus provide actors with a wide scope of possibilities.

Second, *congruence* between the objectives of the core regime and the objectives of functionally linked regimes is likely to increase goal attainment. Overlapping commitments will enhance the authoritative force of international regulation. Different types of overlapping institutions can also fulfil different functions within the same issue area, functions necessary to make an international regime effective. For example, 'soft law' institutions create swift decision-making, but low authoritative force concerning compliance, while 'hard law' institutions tend to be slower on action, but relatively stronger on compliance (Skjærseth, 2003a). It should be borne in mind, however, that overlapping commitments can also lead to duplication of work and low administrative efficiency. For example, different but functionally linked regimes can require separate reporting and verification procedures that drain administrative capacity.

When regimes are formed, the choice of institutional arenas can have important consequences for later events. According to Young (2002, p.121), 'these formative links can and often do set the stage for the emergence of patterns of interplay whose effects are felt long after regime formation is complete.' Likewise, regime linkages can affect international and domestic goal attainment differently to the extent that the latter follows the former. It is worth noting that functionally linked regimes that facilitate goal attainment internationally will not necessarily facilitate goal attainment domestically. For example, the ozone regime has been widely regarded as an innovative institutional model for the climate regime at an early stage. But the implementation of the ozone regime can cause problems for climate-related goals, since some substitutes for ozone-depleting substances also are powerful greenhouse gases. Conversely, linked regimes that do not facilitate goal attainment internationally may serve to further goal attainment domestically. New regimes may emerge and old regimes may evolve, influencing domestic implementation after the positions of states have been initially fixed in core regimes.

Links between international institutions can also occur vertically between the supranational EU and international regimes. The EU does differ considerably from international regimes in scope, depth, and nature, but EU environmental legislation in specific issue areas faces many of the same implementation challenges as international environmental regimes (Skjærseth and Wettestad, 2002). As pointed out by Gehring and Oberthür (forthcoming), whenever a new EU Directive is adopted, it enters an institutional setting that is already densely populated by international environmental treaties. However, EU Directives are also to be implemented domestically in a perhaps even more densely domestic institutional context populated by a high number of existing regulations and other policy instruments. Let us note a rare instance of linked international commitments that applies to Norway: In 1994, Norway joined the European Economic Area (EEA) Agreement with the EU, making about 80 to 90 per cent of EU environmental legislation applicable to Norway even though Norway is not a member of the EU, and does not participate in the main decision-making bodies in the EU system. It has thus been argued that Norway has limited influence on EU environmental decision-making in the form of environmental directives and regulations (Dahl, 1999). The EEA Agreement provides us with an excellent opportunity to explore how overlapping and binding commitments affect goal attainment in a situation where a specific country does not participate in the decision-making body of the linked institution. We can assume that, in cases where access to functionally linked institutions is limited, there will be a high probability for conflict between linked regimes and national environmental goals, and that this situation will tend to reduce the likelihood of domestic goal attainment.

An Alternative Approach: Problem Types

In addition to institutional qualities at the international and national levels, variation in goal attainment is likely to depend upon the characteristics of the problem itself. Goal attainment can be easier in some issue areas than in others, simply because the problem to be solved is more politically 'benign'. Previous studies have shown the type of problem to be an empirically potent variable in the study of the effectiveness of international regimes as well as domestic environmental policy (Skjærseth, 2000; Miles et al., 2002).

Problem types have been variously conceptualised and operationalised (see e.g Rittberger, 1993; Miles et al., 2002). One approach distinguishes problem structures according to actor interests and values as issue-area properties. Another is more concerned with the strategic 'games' generated by different problems. Analytically, both these approaches distinguish between problem types and the institutional rules and procedures applied to solve them. Since the focus in this study is goal attainment by *one state* rather than the level of effectiveness achieved by *one regime*, we will have to modify available indicators that distinguish between 'benign' and 'malign' problems. We follow the issue-area approach, but begin our discussion by relating goal attainment to the relative location of target groups. Four categories can be discerned analytically:

		Outside state: 'outsiders' H	L
Inside state: 'nationals'	H	(1)	(2)
	L	(3)	(4)

Figure 2.1 Location of target groups

1. Balanced location of target groups inside and outside one specific country. Goal attainment here will depend on successful foreign policy capable of strengthening relevant regimes and adequate domestic policy resulting in change in the behaviour of domestic target groups.
2. Most important target groups located inside a country. Goal attainment depends on change in the behaviour of domestic target groups. Failure may occur along this dimension when affected target groups oppose stated goals.
3. Most relevant target groups located outside a country. Goal attainment here will depend on successful foreign policy aimed at affecting foreign target groups by strengthening the rules and procedures of relévant regimes. Failure may occur along this dimension whenever other states in the core regime oppose stated positions.
4. Target groups located neither inside nor outside a country. Goal attainment will to a limited extent depend on change in the behaviour of foreign or national target groups. This situation can occur whenever national goals are similar or close to the actual state of affairs at the time the goals were formulated. This situation parallels coordination problems characterised by fully compatible interests that nevertheless require coordination.

The relative location of target groups cannot in itself capture the degree of political resistance that has to be overcome to achieve environmental goals. We therefore have to supplement location with the compatibility of interests present in the 'upstream' and 'downstream' processes of goal attainment. Interests are closely related to capabilities, power and influence. In the upstream process, the influence of one state in the core regime refers to the control it has over events important to that state. Another face of power refers to actors with the capacity to impose their will upon others. This type of power can be seen as either benevolent or coercive hegemony (Snidal, 1985). The point here is that a state's control over important events will depend on the extent to which its positions in international environmental cooperation are in line with, or in opposition to, the positions of other and perhaps more influential states. The more influential its opponents, the less likely a state will be able to attain its goals internationally. Conversely, the more influential its allies, the more likely a state will be able to attain its goals internationally as part of a 'winning coalition'. In the context of this book, we are particularly interested in the extent to which the involvement of the USA and the EU (the EU here understood as an actor in international environmental cooperation) make any systematic difference for Norwegian goal attainment.[29] Even though a federal state is not directly comparable with a supranational actor, these actors possess significant administrative and knowledge-based capacity, and they both have the capacity to exercise both 'faces' of power in international environmental regimes. The USA and the EU are parties to five of the six regimes selected in this study.

In the downstream process, domestic goal attainment is likely to depend upon the size and distribution of costs and benefits among domestic target groups and other affected actors. In general, strong opposition to national goals can be expected if the bulk of the costs are borne by a few specific target groups while the benefits are widely distributed throughout society. Under such circumstances, implementation becomes extremely difficult. Conversely, strong support of national goals can be expected if the benefits are concentrated within specific target groups whereas the costs are widely distributed (Wilson, 1973; Underdal, 1995).

The upshot of this discussion is that problems caused by target groups inside and outside a given state – as seen from the perspective of that state – are most 'malign' and the prospects for goal attainment most bleak in situations characterised by opposing and influential member-states in the core regime and the presence of opposing target groups at home. Conversely, problems are most 'benign' and the prospects for goal attainment most promising in situations characterised by compatible interests with influential states abroad and supporting target groups at home.

Conclusion: Conditions for Goal Attainment

This chapter has presented the analytical framework that will be used throughout the book for studying variation in Norwegian goal attainment in different environmental issue areas. Since most environmental problems have an element of interdependence rooted in transnational environmental consequences, most states will have to combine both international and domestic efforts to attain their environmental goals. We thus assume that the extent to which Norway is able to attain its goals depends on the extent to which it is able to affect target group behaviour at home and abroad.

We primarily adopt an institutional approach to our inquiry; that is, we focus on domestic institutions, international core regimes, and linked regimes. The reason for our interest in institutional factors is that institutions to varying degrees can be manipulated by decision-makers. However, we also supplement this approach with an alternative that looks at problem types.

Whereas domestic institutions, core regimes and linked regimes are important for both national and international goal attainment, the specific functions that need to be fulfilled internationally and domestically tend to differ as seen from the perspective of the individual state. The explanatory perspectives comprising our institutional approach have thus been related to different institutional factors likely to affect goal attainment internationally and domestically. These factors serve as conditions for goal attainment. Below, the conditions identified in this chapter are summarised accordingly, i.e. across institutions at different levels. The conditions outlined below should not be viewed as exhaustive, but rather as a starting point for exploring why Norway's ability to attain its goals may vary across different issue areas. Notice that some factors may serve purposes both internationally and domestically.

Table 2.1 Conditions affecting Norway's international goal attainment

Independent variables	High ability to attain goals	Low ability to attain goals
Domestic institution	Few domestic branches involved	Many domestic branches involved
	One ministry in charge	Shared competence
Core regime	High activity/leadership exercised	Low activity/leadership not exercised
	Consensus required	Majority voting or unanimity[a]
	Strong regime[b]	Weak regime
Linked regimes	High degree of congruence between regime objectives	Low degree of congruence between regime objectives
	Good opportunities for 'venue shopping'	Poor opportunities for 'venue shopping'
Problem type	Compatible interests with influential actors in core regime	Diverging interests with influential actors in regime

[a] The impact of different decision rules on goal attainment will depend on whether the position taken by a specific state belongs to the majority or minority.

[b] Effect sensitive to congruence between positions and regime objectives.

Table 2.2 Conditions affecting Norway's domestic goal attainment

Independent variables	High ability to attain goals	Low ability to attain goals
Core regime	Strong core regime and congruence between regime objectives and national goals	Strong core regime and divergence between regime objectives and national goals
Linked regime	High degree of congruence between regime objectives	Low degree of congruence between regime objectives
	Access granted to functionally linked regimes	No access to functionally linked regimes
Domestic institution	Sector integration not needed (one domestic sector affected)	Sector integration needed (many domestic sectors affected)
	Permits possible (point sources)	Broad-based portfolio of policy instruments needed (diffuse, or mixed sources)
	Inclusion of those actors responsible for implementation at an early stage	Exclusion of those actors responsible for implementation at an early stage
Problem type	Concentrated benefits and widespread costs: supporting target groups	Concentrated costs and widespread benefits: opposing target groups

The analytical framework outlined in this chapter will guide the empirical analysis in the following chapters. The study follows the case-study approach (Yin, 1989). In Chapter 9, we shall explore the assumptions and conditions in Tables 2.1 and 2.2 across the cases by means of pattern-matching based on narratives. Data collection has been based on interviews, document analysis and secondary sources. In the subsequent chapters we present the following six case studies focusing on Norwegian goal attainment in the following issue areas covered by international regimes:

- Air pollution. Core regime: Convention on Long-Range Transboundary Air Pollution (CLRTAP).
- Marine pollution. Core regime: North Sea Declarations and conventions for the Protection of the Marine Environment of the North-East Atlantic (OSCOM, PARCOM, OSPAR).
- Climate change. Core regime: United Nations Framework Convention on Climate Change (UNFCCC)/Kyoto Protocol.
- Biological diversity. Core regime: Convention on Biological Diversity (CBD).
- Management of whales. Core regime: The International Whaling Commission (IWC).

- Ozone depletion. Core regime: The Vienna Convention and the Montreal Protocol.

These six cases – all focusing on the management of natural resources and the environment – represent a selection from a wider universe of environmental issue areas and international regimes in which Norway is engaged. Our selection is based on three criteria. First, all regimes are politically and/or economically important to Norway and many other states. This criterion is important to minimise the risk of studying environmental goals that were never seriously intended. Second, these cases have been selected to provide significant variance in goal attainment and explanatory factors. Third, pragmatic considerations related to prior knowledge have been one important concern. Case study authors have published extensively on regime effectiveness within their particular fields.

Notes

1 Note the parallel to regime effectiveness studies: the criterion of effectiveness (often generally defined as problem-solving) must be carefully specified before examining and evaluating the behavioural adaptation that can be causally connected to the regime.

2 This line of reasoning will not hold if the relative level of public environmental attitudes changes significantly between different issue areas after goals have been adopted. This possibility cannot be excluded, but the significant drop in environmental concerns in Norway in the mid-1990s appeared mainly as a general trend across various issue areas.

3 Major criticism has been raised against studies of public policy implementation that have been evaluated after only a few years. A time-lag of approximately 10-15 years from the adoption of goals to evaluation is recommended (Sabatier, 1986).

4 This was formulated as an explicit Norwegian goal, but conditioned by various factors and used to facilitate international climate commitments.

5 However, this approach will only partly solve the problem related to the climate target. On the one hand, Norway has ratified the UNFCCC, which aims at limiting man-made greenhouse-gas emissions at their 1990 level in the year 2000. On the other hand, the UNFCCC does not contain legally binding commitments to reduce GHG emissions; it merely expresses an intention.

6 Under the rubric of 'new institutionalism', the study of institutions has developed within various social science disciplines, including economics, sociology and political science; and several relatively distinct schools of thought have developed under this heading.

7 It is important to look at how social demand is channelled through parliamentary and corporate systems in order to understand differences and change in goal attainment in comparative studies of environmental politics (Skjærseth, 1999). However, the issue of social demand is less important when we focus on the attainment of different goals adopted at roughly the same time by the same state.

8 Leadership approaches applied to key persons would clearly represent an alternative and promising perspective for understanding Norwegian environmental foreign policy. In this study, we only hint at the central role played by e.g. Gro Harlem Brundtland and other important officials. However, we have not had the resources available to conduct a systematic in-depth inquiry of leadership at the elite level in Norway.

[9] Notice that we need adapt this approach to foreign policy at the sector level in order to analyse foreign environmental policy within one state.

[10] The Domestic Politics model is also used to explain implementation and compliance.

[11] This aphorism represents the core of Allison's bureaucratic politics model. The argument is that senior advisors representing various agencies will bargain over foreign policy options and present 'stereotypical' advice to decision-makers.

[12] For a study of the implementation of international environmental agreements in Russia with emphasis on regulatory comptence, see Hønneland and Jørgensen, 2003.

[13] For a comprehensive discussion of voluntary agreements, see OECD (2003).

[14] Other relevant criteria include cost-effectiveness, technological innovation potential, transparency, and political feasibility.

[15] Diffuse sources can be regulated by e.g. taxes that tend to score high on cost-effectivenes, but low on goal attainment.

[16] This conclusion is based on a survey of 2,100 companies.

[17] In addition to treaty negotiations, international regimes may come into existence by two other processes (Young, 1989). First, they may be self-generating or spontaneous, i.e. the product of action by many, but not the result of human design. Second, they may be imposed by dominant actors in a process typically not involving consent on the part of subordinate actors.

[18] These variables are distinguished by level of analysis as well as by the behaviour, capabilities, cognition, values and interests, and constitution of the units at different levels of analysis.

[19] Malnes (1995) takes a critical look at the work of Young and Underdal and argues that neither of them distinguishes clearly between the activity of a leader and that of an agent who engages in ordinary bargaining. We apply a lenient interpretation of leadership that is not reserved to an extraordinary role in bargaining.

[20] By 'differential obligations' we mean simply that there are institutional procedures to treat actors differently according to special circumstances. Differential treatment may follow three courses of action. First, there may be a requirement to adopt the same goal, but not the same measures. Second, procedures may open for adopting different goals. Third, there may be a possibility of following different timetables. Regionalisation can be used to restrict membership and lift the common denominator. Regionalisation can be accomplished by splitting up negotiations into sub-processes, for example by establishing sub-institutions to deal with a limited geographical area within the same issue area. This may lead to a de-coupling of the least ambitious parties, and to more stringent international commitments.

[21] A causal approach would require a systematic assessment of regime effectiveness.

[22] Sometimes, one and the same state respond differently to different issues covered by the same regime, even though incentives to defect are present and constant. The UK, for example, chose to defect from the North Sea obligations on nutrients, while complying with obligations on dumping and hazardous substances (Skjærseth, 2003b).

[23] An alternative perspective is based on the economists 'law of diminishing returns'. According to this perspective, the first steps are likely to be the easy ones in which marginal benefits clearly exceed marginal costs. When attempts are made to tighten up joint commitments, marginal abatement costs tend to increase and benefits will tend to decrease. This perspective would suggest that it will become increasingly difficult to step up joint commitments. This scenario is most likely when the original institutional arrangements have a wide scope and include stringent commmitments as well as dynamic qualities.

[24] Recall that if the level of national ambitions is set 'below' the level of international obligations to which Norway has approved voluntarily, the evaluation uses international commitments as an indicator of what can be achieved.

[25] A third pathway suggested in the literature is that international institutional procedures may become enmeshed domestically through the standard operating procedures of bureaucratic agencies (Young, 1989). International institutions may affect implementation and compliance by affecting the relationship between governmental agencies, target groups and ENGOs.

[26] The first systematic empirical assessment of interaction between international institutions (as well as between international institutions and EU directives) will soon be published (Gehring and Oberthür, forthcoming).

[27] The 'proof of the pudding' may in fact be found at the level of non-state target groups, e.g., refineries and oil companies and how they deal with the high number of international and national obligations.

[28] Embedded regimes are deeply embedded in overarching institutional principles and practices, nested regimes are incorporated in broader institutions in the same issue area, and clustered regimes are knit together in more comprehensive institutional packages.

[29] The EU participates in over 30 major international environmental agreements and can be understood both as an actor in international cooperation as well as a sub-regime with regard to its member-states.

References

Aardal, B. (1993), *Energi og Miljø: Nye Stridsspørmål i Møte med Gamle Strukturer*, Institute for Social Research, Oslo, Report 93, p. 15.

Aardal, B. and Valen, H. (1995), *Konflikt og opinion*, NKS-forlaget, Oslo.

Allison, G.T. (1971), *Essence of Decision: Explaining the Cuban Missile Crisis*, Little, Brown, Boston.

Andersen, R. (2002), 'The Time Dimension in International Regime Interplay', *Global Environmental Politics* vol. 2, nr. 3, pp. 98-117.

Andresen, S., Skodvin, T. Underdal, A. and Wettestad, J. (2000), *Science and Politics in International Environmental Regimes: Between Integrity and Involvement*, Manchester University Press, Manchester.

Barkdull, J. and Harris, P.G. (2002), 'Environmental Change and Foreign Policy: A Survey of Theory', *Global Environmental Politics*, vol. 2, no. 2, pp. 63-92.

Chayes, A. and Chayes, A.H. (1993), 'On compliance', *International Organization*, vol. 47, no. 2, Spring, pp. 175-205.

Cortell, A.P. and Davis, J.W. Jr. (1996), 'How Do International Institutions Matter? The Domestic Impact of International Rules and Norms', *International Studies Quarterly*, vol. 40, pp. 451-78.

Dahl, A. (1999), 'Miljøpolitikk – Full Tilpasning Uten Politisk Debatt', in D.H. Claes and B.S. Tranøy (eds), *Utenfor, Annerledes og Suveren? Norge Under EØS-avtalen*, Fagbokforlaget, Bergen, pp. 127-49.

Downs, A. (1972), 'Up and Down with Ecology – The Issue-Attention Cycle', *Public Interest*, vol. 28, Summer, pp. 38-50.

Gehring, T. and Oberthür, S. (forthcoming), *Institutional Interaction – How to Prevent Conflicts and Enhance Synergies Between International and European Environmental Institutions*. Ecologic, Berlin.

Grant, W., Matthews, D. and Newell, P. (2000), *The Effectiveness of European Union Environmental Policy*, McMillian, London.

Grieco, J.M. (1988), 'Anarchy and the Limits of Cooperation: A Realist Critique of the Newest Liberal Institutionalism', *International Organization*, vol. 42, no. 3, Summer, pp. 485-507.

Hall, P.A. and Taylor, R.C.R. (1994), 'Political Science and the Four New Institutionalisms', Paper prepared for presentation to the Annual Meeting of the American Political Science Association, New York, Center for European Studies, Harvard University, Cambridge.

Hønneland, G. and Jørgensen, A.K. (2003), *Implementing International Agreements in Russia*, Manchester University Press, Manchester.

Jänicke, M. (1992), 'Conditions for Environmental Policy Success: An International Comparison', in M. Jachtenfuchs and M.S. Strübel (eds), *Environmental Policy in Europe: Assessment, Challenges and Perspectives*, Nomos, Baden-Baden, pp. 71-97.

Jänicke, M. (1997), 'The Political System's Capacity for Environmental Policy', in M. Jänicke and H. Weidner (eds, in collaboration with H. Jörgens), *National Environmental Policies: A Comparative Study of Capacity-Building*, Springer, Berlin, pp. 1-24.

Jansen, A.I. and Osland, O. (1996), 'Norway', in P.M. Christensen (ed.), *Governing the Environment: Politics, Policy, and Organization in the Nordic Countries*, Nord:5. Nordic Council of Ministers, Copenhagen, pp. 181-259.

Jordan, A.J. (1999), 'The Implementation of EC Environmental Policy: A Policy Problem Without a Political Solution?' *Environment and Planning C (Government and Policy)*, vol. 17, no.1, pp. 69-90.

King, L.A. (1997), 'Institutional Interplay – Research Questions', paper commissioned by Institutional Dimensions of Global Change and International Human Dimensions Programme on Global Environmental Change, Environmental Studies Programme, commissioned by University of Vermont, School of Natural Resources, Burlington.

Levy, M.A., Young, O.R and Zürn M. (1995), 'The Study of International Regimes', *European Journal of International Relations*, vol. 1, no. 3, pp. 267-330.

Malnes, R. (1995), '"Leader" and "Entrepreneur" in International Negotiations: A Conceptual Analysis', *European Journal of International Relations*, vol. 1, no.1, pp. 87-112.

Mearsheimers, J. (1995), 'The False Promise of International Institutions', *International Security*, vol. 19, nr. 3, Winter 1994/95, pp. 5-49.

Miles, E.L., Underdal, A., Andresen, S., Wettestad, J., Skjærseth, J.B. and Carlin, E.M. (2002), *Environmental Regime Effectiveness: Confronting Theory with Evidence*, The MIT Press, Cambridge, MA.

Ministry of the Environment (1989), *Environment and Development. Programme for Norway's Follow-up of the Report of the World Commission on Environment and Development*, (Miljø og Utvikling. Norges oppfølging av Verdenskommisjonens rapport), Report No. 46 to the Storting (1988-89), Ministry of the Environment, Oslo.

Nollkaemper, A. (1993), *The Legal Regime for Transboundary Water Pollution: Between Discretion and Constraint*, Graham & Trotman, London.

OECD (2003), *Voluntary Approaches for Environmental Policy: Effectiveness, Efficiency and Usage in Policy Mixes*, OECD, Paris.

Ostrom, E. (1990), *Governing the Commons: The Evolution of Institutions for Collective Action*, Cambridge University Press, Cambridge.

Powell, W.W. and DiMaggio, P.J. (eds) (1991), *The New Institutionalism in Organizational Analysis*, The University of Chicago Press, Chicago, and London.

Rittberger, V. (ed. with P. Mayer) (1993), *Regime Theory and International Relations*, Clarendon Press, Oxford.

Rosendal, K. (2001), 'Impacts of Overlapping International Regimes: The Case of Biodiversity', *Global Governance*, vol. 7, pp. 95-117.

Ruggie, J.G. (1983), 'International Regimes, Transactions, and Change: Embedded Liberalism in the Postwar Economic Order', in S.D. Krasner (ed.), *International Regimes*, Cornell University Press, Ithaca and London, pp. 195-231.

Sabatier, P. (1986), 'Top-Down and Bottom-Up Approaches to Implementation Research: A Critical Analysis and Suggested Synthesis', *Journal of Public Policy*, vol. 6, no. 1, pp. 21-48.

Sand, P.H. (1991), 'Lessons Learned in Global Environmental Governance', *Environmental Affairs Law Review*, vol. 18, pp. 213-77.

Skjærseth, J.B. and Rosendal, G.K. (eds) (1995), 'Norwegian Environmental Foreign Policy', in T.L. Knutsen, G.M. Sørbø and S. Gjerdåker, *Norwegian Foreign Policy* (Norges utenrikspolitikk), Cappelen, Oslo, pp. 161-80.

Skjærseth, J.B. (1999), *The Making and Implementation of North Sea Pollution Commitments: Institutions, Rationality and Norms*, Akademika AS and the Faculty of Social Sciences, University of Oslo.

Skjærseth, J.B. (2000), *North Sea Cooperation: Linking International and Domestic Pollution Control*, Manchester University Press, Manchester.

Skjærseth, J.B. and Wettestad, J. (2002), 'Understanding the Effectiveness of EU Environmental Policy: How Can Regime Analysis Contribute?' *Environmental Politics*, vol. 11, no. 3, pp. 99-120.

Skjærseth, J.B. (2003a), 'Protecting the North-East Atlantic: Enhancing Synergies by Institutional Design', paper presented at the 44th Annual ISA Convention, Portland, Oregon, February 26 to March 1, 2003.

Skjærseth, J.B. (2003b), 'Managing North Sea Pollution Effectively: Linking International and Domestic Institutions', *International Environmental Agreements: Politics, Law and Economics*, vol. 3, pp. 167-90.

Snidal, D. (1985), 'The Limits of Hegemonic Stability Theory', *International Organization*, vol. 39, no. 4, Autumn, pp. 579-614.

Stokke, O. (2000), Managing Straddling Stocks: 'The Interplay of Global and Regional Regimes', *Ocean & Coastal Management*, vol. 43, pp. 205-34.

Stokke, O.S and Vidas, D. (1996), 'Effectiveness and Legitimacy of International Regimes', in O.S. Stokke and D. Vidas (eds), *Govering the Antarctic: The Effectiveness and Legitimacy of the Antarctic Treaty System*, Cambridge University Press, Cambridge, pp. 13-31.

Tallberg, J. (2002), 'Paths to Compliance: Enforcement, Management, and the European Union', *International Organization*, vol. 56, nr. 3, pp. 609-43.

Underdal, A. (1979), 'Issues Determine Politics Determine Policies: The Case for a "Rationalistic" Approach to the Study of Foreign Policy Decision-Making', *Cooperation and Conflict*, vol. 14, no. 1, pp. 1-9.

Underdal, A. (1980a), *The Politics of International Fisheries Management: The Case of the Northeast Atlantic*, Universitetsforlaget, Oslo.

Underdal, A. (1980b), *'Integrated' Marine Policy: What? Why? How?*, R:004, The Fridtjof Nansen Institute, Lysaker.

Underdal, A. (1991), 'Solving Collective Problems: Notes on Three Modes of Leadership', in *Callenges of a Changing World, Festschrift to Willy Østreng*, The Fridtjof Nansen Institute, Lysaker.

Underdal, A. (1995), 'Implementing International Environmental Accords: Explaining "Success" and "Failure"', paper prepared for presentation at the 36th Annual

Convention of the International Studies Association, 21-25 February, Chicago. Departement of Political Science, University of Oslo, Oslo.

Underdal, A. and Hanf, K. (2000), 'International Environmental Agreements and Domestic Politics', Ashgate, Aldershot.

Vedung, E. (1997), *Policy Instruments: Typologies and Theories*, Uppsala University, Department of Government, Uppsala.

Waltz, K.N. (1979), *Theory of International Politics*, Addison-Wesley, Reading, MA.

Weale, A. (1992), *The New Politics of Pollution*, Manchester University Press, Manchester.

Weale, A. and Williams, A. (1993), 'Between Economy and Ecology? The Single Market and the Intergration of Environmental Policy', in D. Judge (ed.), *A Green Dimension for the European Community: Political Issues and Processes*, Frank Cass, London, pp. 45-64.

Weiler, J.H.H. (1991), 'The Transformation of Europe', *The Yale Law Journal*, vol. 100, pp. 2401-83.

Weiss, E.B. (1993), 'International Environmental Issues and the Emergence of a New World Order', *Georgetown Law Journal*, vol. 81, no. 3, pp. 675-710.

Wettestad, J. (1999), *Designing Effective Environmental Regimes: The Key Conditions*, Edward Elgar, Aldershot.

Wilkinson, D. (1997), 'Towards Sustainability in the European Union? Steps Within the European Commission Towards Integrating the Environment into Other European Union Policy Sectors', *Environmental Politics*, vol. 6, nr. 1, pp. 153-73.

Wilson, J.Q. (1973), *Political Organizations*, Basic Books, New York.

World Commission on Environment and Development (Verdenskommisjonen for miljø og utvikling) (1987), *Our Common Future* (Vår felles framtid), Tiden Norsk Forlag, Oslo.

Yin, R.K. (1989), *Case Study Research: Design and Methods*, Sage, London.

Young, O.R. (1979), *Compliance and Public Authority. A Theory with International Applications*, Johns Hopkins University Press, Baltimore, MD and London.

Young, O.R. (1986), 'International Regimes: Toward a New Theory of Institutions', *World Politics*, vol. 39, no. 1, October, pp. 104-23.

Young, O.R. (1988), 'The Politics of International Regime Formation: Managing Natural Resources and the Environment', *International Organization*, vol. 43, no. 3, 361-2.

Young, O.R. (1989), *International Cooperation. Building Regimes for Natural Resources and the Environment*, Cornell University Press, Ithaca, NY.

Young, O.R. (1991), 'Political Leadership and Regime Formation: On the Development of Institutions in International Society', *International Organization*, vol. 45, no. 3, Summer, 281-308.

Young, O.R. (1994), *International Governance: Protecting the Environment in a Stateless Society*, Cornell University Press, Itaca, NY and London.

Young, O.R. (1996), 'Institutional Linkages in International Society: Polar Perspectives', *Global Governance*, vol. 2, no. 1, pp. 1-24.

Young, O.R. (2002), *The Institutional Dimensions of Environmental Change: Fit, Interplay and Scale*, The MIT Press, Cambridge, M.A.

Zürn, M. (1991), 'Consequences of Regime Definitions and Definitions of Regime Consequences: Proposals for a Data Bank on International Regimes', working paper presented at the workshop entitled 'Regimes Summit', Institute of Arctic Studies, Dartmouth College, Hanover, New Hampshire, October.

Chapter 3

Whaling: Peace at Home, War Abroad

Steinar Andresen

Introduction

Norwegian whaling policy has been one of the few areas where Norway has been at odds with all its traditional allies. Over the last two decades Norway has insisted on continuing to hunt whales commercially despite the international ban on the practice. Thereby the whaling issue stands out as a rare case in Norway's traditionally peaceful but also ambitious broker role on the international environmental scene. Domestically, however, all key actors have rallied in support of the official Norwegian whaling policy.

Historically, Norway has been a major industrial whaling nation and was a pioneer on the international whaling scene until World War II. Thereafter Norway's role gradually declined, and from the 1970s only small-scale coastal whaling was conducted. After pressure from anti-whaling forces, Norway quit commercial whaling in the mid 1980s, but it was resumed in the early 1990s. In 2001 Norway also opened up for the export of whale products.

The whaling case differs in several noteworthy ways from the other issue areas discussed in this book. First, although whaling by many countries is perceived as an environmental issue, Norwegian decision makers have always regarded this as a question of managing marine living resources – explaining why the whaling issue is not included in the White Paper on the follow up of the Brundtland Commission (Ministry of the Environment, 1989). Second, the game of implementation is not relevant in this case; it is a story about policy making and decision making. Third, while the other issues covered in this book have a fairly short history – ranging from two to three decades – the peak in the global catch of whales was some seventy years ago.

By first tracing the background of whaling, this chapter seeks to explore why Norway has taken such a strong stance on the whaling case, despite international pressure. It looks into the degree to which Norway has attained its goals at both international and domestic levels. It is argued that a high level political strategy is the main reason that all major domestic players rallied behind the position to resume whaling. Some success has been achieved internationally as well, but opposition has continued to be strong, although somewhat reduced over time as the whaling issue is no longer as 'hot' politically as it used to be.

Industrial Commercial Whaling Depletion, Failed Regulation and the Role of Norway[1]

Historically, whaling has been a major Norwegian industry and one of the few areas, before the area of oil and gas, where Norway was a major actor on the international scene. Traditionally, whaling was a trade that demanded a pioneer spirit and daring adventure; skills, bravery, technological innovation, shipping and hunting were combined in such a way that Norway became a world whaling power. In fact, modern whaling was introduced by Norway around 1870 outside the coast of its most northern county, Finnmark. From there it expanded to every ocean area of the world where large whales could be hunted, and all major sea-going nations were also whale hunters. Profit was large, and the whaling industry grew – especially when whaling moved to the Southern Ocean around 1900 where major whale species were extremely abundant. The whales were hunted exclusively for their oil,[2] and the peak in production was around 1930. At the time Norway and the UK accounted for some 90 per cent of the whales caught. After World War II the Soviet Union and Japan gradually became the major whaling nations.[3] Catch continued at a high level until the early 1960s, thereafter declining sharply, with commercial catches in Antarctic waters ceasing in the mid 1980s.

In short, this industrial adventure lasted for about a hundred years, and it is one of the bleakest stories of resource depletion the world has ever seen. For example, whalers used to take thousands of blue whales per season, but Japan did not agree to quit commercial whaling of blue whales until numbers had been so depleted that their catch was down to 21 animals in one season.

Various attempts to regulate whaling internationally were introduced in the 1930s, but with little effect. In 1948 the International Whaling Commission (IWC) was established, but was no more successful than its predecessors until the mid 1960s when the effectiveness of the organization gradually increased (Andresen, 2001b). At the time, the question of international whale management was a curiosity in the fringes of international politics. This changed abruptly when the whale was adopted as a symbol of the international environmental movement and when at the 1972 UN Stockholm Conference a resolution for a ten-year moratorium against commercial whaling was unanimously adopted.[4] It took some time before the anti-whaling sentiment overtook the IWC, but large-scale industrial whaling gradually faded out.

Norway has been characterized as a pioneer in adopting national whaling regulations. Nevertheless, up until World War II the *market mechanism* through supply and demand (for whale oil) was by far the most important regulator. In the 1930s Norway, and increasingly the UK, were worried about shrinking demand and increasing supply and therefore tried to reduce the amount of whale oil through various *production control agreements.* As the aggressive newcomers (Japan and Germany) did not share this concern, they were not very successful. In short, prior to World War II the concern for reducing catch was only motivated by economic concerns, not the need to conserve whales as a means in itself, but competition between key actors reduced the effectiveness of the agreements.

Norway was a key player in establishing the IWC, and its position was strengthened by the fact that Japan and Germany were not allowed to take part.[5] Norway had the most significant expertise with regard to manpower and technical skills as well as scientific knowledge. It was, however, not Norway that stood forth as the major player on the international whaling scene, but the UK and particularly the US. When it came to international politics in the early post-war period, the US was the dominant power in all kinds of international cooperation, including the establishment of the IWC.[6] There were, however, no major controversies in the making of the whaling convention. It was considered novel at the time, as it tried to strike a balance between conservation and utilization.

As a major player on the scene for some 80 years, Norway's main goal was to keep its position within the framework of the newly established IWC, but was not at all successful on this account.[7] The old whaling 'giants' were overrun by the two aggressive newcomers. These newcomers had newer and more advanced technology and equipment, and the governments of Japan and the Soviet Union also actively backed their expanding whaling fleets. Thus, by the end of the 1960s the Norwegian industrial whaling fleet was wiped out. In the IWC Scientific Committee, Norwegian scientists worked actively with other scientists to reduce quotas, but with little effect (Schweder, 2000, Andresen, 2000). Norway was still an important player due to its previous position, knowledge and expertise, but its influence was reduced due to its shrinking whale fleet and thereby its interest and concern for whaling regulations. Norwegian Antarctic operations could not be maintained, competition was too hard, and resources were too scarce in supply; Southern Ocean whaling was simply not profitable any more.

In short, around 1970 Norway stood forth as a knowledgeable player, but was stripped of its Antarctic whaling fleet. In contrast to the pre-World War II period, a stable international regime existed, but Norway had not been able to shape its direction in a way that protected Norwegian interests – or the whales.

More Aggressive Whaling Policy: Increased Goal Attainment?

Although Norway quit Antarctic whaling operations, this did not mean it lost all interest in the whaling issue. As will be described in more detail later, coastal whaling continued. Therefore, the Norwegian goal was to continue to catch these whales commercially.

As a result of the increased political attention and polarization concerning the whaling issue, a number of non-whaling nations joined the IWC, and in 1982 a moratorium against commercial whaling was adopted with a three-quarter majority.[8] In short, while the IWC started out as a 'whalers club', after the adoption of the 1982 moratorium it gradually turned into a 'protectionist club' (Andresen, 2001a). That is, while the IWC used to be extremely unsuccessful in conserving whales, over the last two decades the IWC has been highly successful on this account. Large-scale industrial whaling has ended, key whale stocks are growing, and some of them have once more become abundant.[9] In this sense, the IWC stands forth as an unusually strong international regime.

The IWC is still the core regime pertaining to whaling, but a host of other relevant international institutions have been established more recently – a reflection of strong growth in international environmental institutions. It has been noted that the 1972 Stockholm Conference had very important consequences for the whaling issue, and it was also discussed at the summits in Rio (1992) and Johannesburg (2002). The most important forum for the management of marine living resources in general, however, was the Third Law of the Sea Conference (UNCLOS III), 1973-82. There are also relevant regional newcomers. The most important body so far is the North Atlantic Marine Mammal Commission (NAMMCO). The General Agreement on Trade and Tariffs (GATT) and World Trade Organization (WTO) rules also have consequences for the management of whaling. The most important 'external' regime for the whaling issue, is the Convention on International Trade in Endangered Species of Wild Flora and Fauna (CITES), which is decisive for the possibilities of exporting whale products. Finally, EU rules pertaining to catch and export of whale products are highly relevant for the pending question of possible Norwegian EU membership. These arenas and institutions will be discussed in more detail later. The following section will first discuss Norway's 'score' in terms of goal attainment.

Norway Quits Whaling: Low Goal Attainment

Norway continued to conduct small-type coastal whaling, mainly minke whales from the North East Atlantic stock, when the industrial 'whaling adventure' ended. No political attention was paid to coastal whaling, as it was seen as a small appendix to Norwegian fishing policy. Moreover, as minke whaling was not regulated by the IWC in the early 1970s, Norway did not have any direct material interests to protect after its Antarctic operations ended. In fact, a previous Norwegian Commissioner to the IWC has described the IWC as more like a 'travel club' for the Norwegian delegation in this period.[10] However, the IWC gradually increased its regulatory scope, and by the mid 1970s all major whale species, including the North East Atlantic minke whale, were regulated by the IWC.

The political pressure on Norway to cease whaling increased with the adoption of the 1982 moratorium. Norway opposed and rejected the moratorium and was therefore not legally bound by it. Also, Norwegian scientists claimed that the catch was within safe scientific limits, and the whalers wanted to continue catching minke whales. Nevertheless, in July 1986 Norway announced that it would halt commercial whaling after the 1987 season. Considering that the goal was to continue whaling commercially, it seemed goal achievement was at an all-time low: A previous whaling giant decides to stop whaling altogether – contrary to its goals and beliefs.

It is necessary, however, to scrutinize the decision to stop commercial whaling more in depth. Norway stipulated that the halt would be *temporary* and that Norway maintains that all marine living resources must be managed on a proper scientific basis. That is, Norway made it clear that it would resume whaling if and when it was scientifically justified, not because it received the blessing of the majority in the IWC. Still, the decision to stop whaling implied that goal achievement was low.

Coastal Whaling is Resumed: Increased Goal Achievement

At the 1992 IWC meeting, Norway, to the surprise of domestic as well as international audiences, declared that it would resume commercial whaling from the 1993 season. This decision was formally confirmed on 18 May 1993, when the Minister of Foreign Affairs, Johan Jørgen Holst, addressed the Norwegian Parliament. Since 1993 Norway has conducted commercial minke whaling on a small scale. The number of minke whales taken per season has increased from some 300 animals in 1993 up to some 600-700 animals over the last few years.

As noted, there is no mention of whaling in the 1989 White Paper. In his address to Parliament in 1993, however, the Foreign Minister for the first time and in more comprehensive terms laid out the broader goals of Norwegian whaling policy by presenting the reasons for resuming commercial whaling (*Forhandlinger i Stortinget*, 1993). He maintained that the scientific element was still important, but that it must be revised in light of new knowledge and concepts introduced in the meantime. First, he argued that limited whaling was justified from a *scientific* perspective because the stock was abundant. Second, he pointed out that for coastal states it was deemed very important to secure an effective *multi-species* management. In essence, the fishermen would be hurt economically if this specific species was protected. Third, 'and perhaps most importantly' (ibid, p.3855, author's translation), he argued that the issue of preserving rural livelihoods was far more important than whaling – that Norway's decision to resume whaling should be seen as the right of a small country to utilize accessible natural resources in a sustainable way. Because the Norwegian economy relies heavily on natural resources, he said, this is *an important principle which we cannot abandon* (Authors emphasis). Fourth, he argued that it was important for small states to *withstand pressure from powerful actors* (author's emphasis) because the public opinion that influenced these powerful actors was based on a disrespect for the rules of international agreements and resource management. He said that small states like Norway depended on mutual respect of international agreements, and that although whaling was not important from a national *economic* point of view, it also represented important vocational opportunities in areas where few alternatives existed.

These political, economic, legal and scientific considerations all weighed heavily in the direction of resuming whaling. Still, the Foreign Minister stated explicitly that there were considerable *risks* associated with this decision. There would be potential ec*onomic* costs of resuming whaling because of threats of economic sanctions from the US, threats of various kinds of consumer boycotts from Greenpeace and other NGOs, and even threats of more violent actions from more extreme groups. These potential costs, however, were not considered sufficient to refrain from resuming whaling. The potential *political* costs in terms of reduced environmental credibility on the part of Norway were not mentioned in this statement. The fact that Norway was seen by important allies as violating 'the spirit' of international law and politics by resuming whaling against the will of the overwhelming IWC majority was not discussed either. However, both these factors were clearly considered by the decision makers.

As long as catch was limited to satisfy domestic consumption only, it was bound to be modest. In this sense it appeared that Norway was escalating the 'whale war' through the decision to resume export of whale products. That is, to resume export was in itself seen as an aggressive move, and it might also contribute to increase the catch of whales.

Resumed Export: Increased Goal Attainment

In 2001 Norway somewhat surprisingly declared that it would resume *export* of whale products after some 15 years of having a self-declared ban on such export. Although the ban was unilateral, it no doubt had bearing upon the fact that all export of whale products had been prohibited by CITES since the mid 1980s. Still, Norway was not *legally* bound by this decision, but *politically* the costs of export may have been seen as too high.

Strict conditions on control and compliance accompanied the decision to resume export. A DNA registry for minke whales was established, and every whale taken by Norwegian whalers was DNA tested to make sure that no illegal export should take place.[11] It took longer time than expected to get the system operative, but it has now been tested by external independent institutions and has passed the test. Moreover, Norwegian authorities require that the importers of these products have the same DNA system in place. Japan was expected to represent the major market for Norwegian whale products, but due to internal strife in Japan, it is still uncertain if and when there will be any export to Japan.[12] However, there have been some small shipments of whale meat to Iceland and the Faroe Islands – insignificant from a practical economic point of view, but important as a question of principle.

Officially the decision to resume export was essentially based on the same arguments used when it was decided to resume commercial whaling. The stock was healthy, there were no legal barriers to export to selected countries, and export would increase utilization of the whales, not the least through export of stockpiled blubber to Japan, as well as increase the profitability of the whaling industry significantly.

If the huge Japanese market is opened up for export it will certainly increase goal attainment in *economic* terms. So far, perhaps the most significant aspect of Norwegian whaling has been the justification of Norway's *right* to whale as a matter of principle; the economic dimension has not been important. At present (2003), there is little likelihood that there will be a strong increase in export. Reactions to the Norwegian escalation of the 'whale war' through opening up for export have been modest, indicating that the whaling issue is no longer as 'hot' as it once was. There is widespread satisfaction with the Norwegian control system, but fears have been expressed that others may not provide similar procedures.

Overall Goal Achievement: Medium

Where does this leave us in terms of goal achievement? No doubt it was a victory for the small state of Norway to resume whaling, against all odds, when this was the declared goal. At the time the decision was made, the potential economic and political costs were unknown. A calculated risk was taken. As it turned out, however, the economic costs have been marginal because various threats of boycott and sanctions never materialized. The decision implied that the trend of yielding unwillingly to the majority was broken. That is, in these terms goal achievement was high.

Goal achievement was lower when it came to influencing the core regime in question, the IWC. Norway was unable to persuade the members of the validity of letting science determine whether or not whaling should be resumed, and was condemned by the large majority. Moreover, many environmentalists as well as key allies saw the whaling issue as a 'blemish' on Norway's environmental record. Over time, however, goal achievement has increased somewhat along this dimension as well. Although there is still a majority against commercial whaling in the IWC, the dominance of the anti-whaling forces has been considerably reduced. Also, much less negative attention is paid today to the fact that Norway is a whaling nation than was the case 10 years ago. On the other hand, the very moderate Norwegian catch is to some extent a concession to these anti-whaling forces. Overall then, a medium score seems appropriate.

The more general Norwegian goal of being an honest broker and true internationalist, however, has been given a blow – but this aspect has also softened over time with the increasing marginalization of the whaling issue.

Explaining Norwegian Goal Attainment

The following discussion will focus primarily on the period from the adoption of the moratorium until present. The *interaction* between the domestic and the international level is the key to understanding the development of Norwegian whaling policy and Norway's goal attainment. Let us start first with the development on the domestic scene. The key issue is how to explain why Norway quit whaling and why it was decided to resume commercial whaling.

The Domestic Scene: The Main Actors: Position, Policies and Influence[13]

Initially, Norwegian small-scale whaling was a simple process, with participation restricted to the fisheries sector. The Ministry of Fisheries and their operating arm, the Fisheries Directorate, were the regulatory agents, and most scientific work was done by the Institute of Marine Research (IMR), also part of the official fisheries bureaucracy. The target group, (fishers and whalers) had access to the policy process through their organizations, the Fishermen's Union, and the Norwegian Small-Type Whalers Association. It was a consensual and closed process; as costs and benefits were perceived to be highly concentrated in the fisheries sector, no other actors needed to interfere.

Increased external pressure on Norway in the mid 1980s to halt commercial whaling triggered an expansion in the type and number of domestic actors involved in the debate. In general this changed the game from one of very 'low politics' to one coming close to 'high politics' – rare in the field of international environmental and resource policy. Formally, the Ministry of Foreign Affairs has always been an important player, as the commissioner usually comes from this ministry. Still, in general the Ministry of Fisheries has been the key governmental actor in the day-to-day business of whaling. The external pressure, including threats of sanctions and consumer boycotts, changed this and shifted whaling from the domain of fisheries to that of foreign policy, making the Ministry of Foreign Affairs a key actor. The Ministry of the Environment was also engaged, as the whaling issue had also been turned into a question of environmental policy. Another new actor on the scene was the Ministry of Trade, as threats of boycotts and sanctions affected wider Norwegian trade interests. At the sub-national level, external pressure triggered responses from the media, trade and industry organizations, green NGOs, scientists, and of course the target groups, fishermen and whalers.

Many of these new actors managed to create the impression that continued whaling could hurt important parts of the Norwegian economy and have considerable economic costs. For example, the Ministry of Foreign Affairs was inclined to focus mostly on the *problems* caused by the Norwegian whaling that could threaten Norway's good reputation. Within a broad foreign policy perspective, to quit whaling on a more permanent basis might seem the only logical conclusion. It may well be that if the decision had been left to the Ministry of Foreign Affairs that the last whale would have been taken by Norway in 1987.[14] In short, the target group was the only direct beneficiary of whaling, and outside the fisheries sector it seemed to have few supporters. How can it then be explained that Norway still resumed commercial whaling?

First, the fisheries sector, including governmental and non-governmental actors, is quite an important political actor in Norwegian politics. It was not self-evident that the whalers would get support from the fishers, as they might have feared losses in income from export revenues if whaling resumed. This was, however, seen as a matter of *principle*, and recall that the whalers were also fishers most of the year. Generally this group saw the IWC as fundamentally flawed; they were tired of being harassed at the yearly IWC sessions and wanted Norway to leave the IWC and instead try to set up some alternative management body.[15] In principle this was certainly an option, but other key actors like the Ministry of Foreign Affairs did not share this view. Norway did not want a reputation for leaving legitimate international bodies just because they did not serve Norwegian interests.

Although the question of Norway leaving the IWC was not a real option, the question of whether or not to resume whaling remained unresolved. The two most important ministries were split on the issue, but another actor intervened: the Prime Minister's Office and Gro Harlem Brundtland, after she resumed her position as

Prime Minister in 1986. Previously, Ms. Brundtland had led the UN Commission on Sustainable Development.[16] She began her career as Minister of the Environment in the 1970s, and throughout her career had been an ardent advocate of scientifically based decision making regarding international resource and environmental issues. She had been following the scientific controversy between Norwegian and other scientists closely in the mid 1980s, and had gotten the impression that Norway's scientific case was weak. This is why whaling was ceased – and a scientific strategy launched. If Norway had a weak scientific basis, it could be very damaging both to her reputation as the 'world minister' of the environment as well as to Norway more generally.

This 1986 decision to halt whaling temporarily pending a thorough scientific investigation of the status of the stock was a carefully thought-out decision, balancing domestic and international demands. Norway was influenced by the demand from the IWC majority to end commercial whaling, but it was done temporarily to reduce the opposition from the vocal fishermen and whalers organizations. Also, it was indirectly admitted that the scientific basis for Norwegian catch was not good enough. As a concession to the domestic fisheries segment, it was made clear that *if* it could be proven beyond doubt that the stock was abundant, harvesting would likely resume. This compromise decision meant that neither domestic whalers nor international anti-whalers were entirely happy, but there was something in it for both groups, and it bought Norwegian decision-makers some more time to consider this thorny issue.

Thereby a large-scale *scientific* strategy was launched. It is important to note how this process was organized. Traditionally, whale-related research had been carried out by the Institute of Marine Research (IMR) in Bergen, which at the time was directly under the Ministry of Fisheries. The polarization of the issue raised the question of the *independence* of science. A group of four scientists was set up, including leading non-Norwegian international scientists. The non-involvement of the IMR underlined the independence of the group and increased its legitimacy. The task of this group was to assess previous Norwegian research on the issue. Their conclusions were probably more critical to previous Norwegian whaling research than Norwegian authorities had expected. One of their main observations was the following: 'Over the years a total of more than 100,000 minke whales have been taken by Norway. However, very little data useful for the management of this species has resulted from this catch.' (Anderson, 1987)

To understand the subsequent direction of Norwegian whaling policy it is important to note the key role played by professor Lars Walløe, University of Oslo, in this process. He had not previously been directly involved in this kind of research, but his research credentials were impeccable. Moreover, he had previously been involved in the dispute between the UK and Norway over the effects of acid rain. It was Ms. Brundtland who had brought him into that issue, in her capacity as Minister of the Environment, and now she brought him into another even more controversial issue.[17] He was a member of the 'group of four' international experts, and he was appointed leader of the large research program launched to learn more about the status of the Northeast Atlantic minke whale stock.[18] Since 1987 and up to the present he has been a key figure both on the international and domestic scientific scene, and has represented the *continuity* of Norwegian whaling policy.

The main part of the program consisted of new methods to count and calculate the size of the stock. For the purposes of the project, 51 whales were taken in the period 1988-91. Research findings indicated that the whale population was *more* abundant than previously thought. To learn more about the role of the minke whale in the marine ecosystem, it was decided in 1991 to escalate the scientific catch considerably, and more than 300 whales were taken in the 1992-94 period. Throughout this period, Norwegian scientists worked in close cooperation with the IWC Scientific Committee. At the IWC meeting in 1991, the Scientific Committee endorsed Norway's research and accepted the Norwegian point estimate of 86,700 animals. This concluded this part of the research program that was part of the scientific strategy. Although the Scientific Committee accepted the Norwegian figures, the political body, the Commission was not convinced. 'Hate-resolutions' were adopted against Norwegian scientific whaling by the IWC majority, and the US certified Norway for sanctions when the scientific catch was as low as five animals a year!

In other words, a scientific strategy was not sufficient; a *political* strategy was needed as well. The political strategy was aimed both at domestic groups and at actors at the international level. As previously noted, most domestic actors with an opinion on or interest in the issue were most inclined to favor the cessation of whaling out of fear of political and/or economic costs of continued whaling. Moreover, this was essentially a 'non-issue' among the people at large, and whale meat was no longer an important part of the diet of most people. In short, the *most likely* scenario seemed to be that Norway would stop whaling.

Still, the Norwegian Prime Minister had decided to rely on the science as a point of departure, and the scientists had spoken: sustainable Norwegian whaling could be conducted. This provided the basis for an *offensive political strategy*; reluctant domestic players needed to be convinced that this was an important matter of principle, irrespective of its marginal economic role. The whalers union and the whaling communities also lobbied actively as well as efficiently in favor of resuming whaling, and the Prime Minister's Office was receptive to their arguments. A small informal group was established to convince domestic and international audiences that it was justified for Norway to resume whaling. The Prime Minister's Office, key people from the ministries of foreign affairs and fisheries, and key scientists took part. Some very able people from the Ministry of Foreign Affairs were put to work on the issue. It was said to be easy to involve these people because the task was considered a real challenge – more interesting than most traditional diplomatic issues. Internationally, lobbying was intense, not the least in the US, both in relation to Congress as well as the administration, but a number of other key countries as well as key environmental organizations were also targeted.[19] Importantly, both Norwegian ministers of foreign affairs in the early 1990s (Thorvald Stoltenberg and Johan Jørgen Holst) were vocal spokesmen in favor of sustainable commercial whaling, both on the domestic as well as at the international scene.

Still, when Norway at the 1992 IWC Commission meeting declared that it would resume commercial whaling from the 1993 season, this came much to the surprise of domestic as well as international audiences. The final decision to resume commercial whaling was not, however, made according to 'standard bureaucratic procedures'. The decision was made by the Prime Minister and the Minister of Foreign Affairs. In fact, members of the ministry's bureaucracy, much to their annoyance, were not even informed that the decision had been made until it was announced publicly. The US delegation, even more annoyed, was also informed only half an hour before the decision was declared.

The key decision-making group was small and exclusive, but a wide range of stakeholders had been consulted prior to the decision. Close links were established also with potential opponents of Norwegian whaling policy, such as the most important Norwegian environmental NGOs, as well as various groups representing Norwegian export interests. These regular contacts were important in building domestic consensus. As noted, prior to these consensus building overtures, parts of the media, parts of the green NGO community, and potentially affected export and business interests wanted to end commercial whaling altogether. These groups were now drawn into the decision making process and opposition gradually waned. As to the role of Parliament, the Extended Committee on Foreign Relations had also been consulted and informed on a regular basis.

This decision was quite unique in the history of Norwegian international relations because Norway, the typical small state internationalist, vocally opposed all major allies – as well as the international green movement. Why was this done? Among one of the key players in the process, it has been described as a 'gut reaction'; it simply was not *right* for outsiders, considered to have no understanding of the issue, to decide this question. Moreover, it represented a rather rare alliance between a small 'periphery' and a small high-level group. The position of the 'small' periphery was probably strengthened somewhat by the ongoing negotiations with the EU on Norwegian membership. The rural districts in Norway and particularly the fishing communities strongly opposed Norwegian membership. The positive attitude by the Government towards whaling may have been an attempt to improve relations with these actors.[20] The fact that the scientific basis was so sound as well as consensual was also crucial. Had this not been the case, whaling would probably have ended.

Gradually all major actors, including all major Norwegian green NGOs, rallied behind the official Norwegian position. When Minister of Foreign Affairs, Johan Jørgen Holst, formally confirmed the decision to resume whaling 18 May 1993 in the Norwegian Parliament, there was not one single vote against this move. No doubt the strong external pressure had been counterproductive in the sense that it tended to *unify* the Norwegian population on the issue and weaken whatever modest opposition there might have been.[21]

After the decision to resume whaling was made, Norwegian whaling policy at the domestic level gradually moved into more smooth and stable waters. As the external threats of sanctions and boycotts did not amount to much, the modest opposition to Norwegian whaling that had remained gradually disappeared. The main actors remained essentially the same – although because it was no longer such a 'high-level' issue, the Prime Minister and the Prime Minister's Office became less involved. The fishing sector and Ministry of Foreign Affairs were the key actors. As there were frequent shifts in personnel working with this issue both in the Ministry of Foreign Affairs and the Ministry of Fisheries, professor Lars Walløe continued to represent the continuity and institutional memory in Norwegian whaling policy.[22]

There was a gradual increase in the quota during the late 1990s until it leveled out at around 600-700 animals. The Whalers Union and not least the spokesman of the newly established Coastal Party, former whaler Steinar Bastesen, was critical of the size of the quota, claiming that the stock could easily support a catch of some 2000 animals. It was claimed that an increased quota was also necessary from an ecological point of view, as the increasing minke whale stock was a threat to various species of fish. No doubt the catch remained moderate in part to keep in line with the very cautious management procedure of the IWC Scientific Committee, although Norway was not bound by it, as the IWC sets no commercial quotas. However, it was also partly because Norway did not want to provoke opponents of Norwegian whaling unnecessarily, and partly because the Norwegian market did not have a demand for more than this level of catch.

This brings us to the question of resuming the export of whale products. The Whalers Union had long argued in favor of opening export as a means to increase profit, catch and the utilization of the whales. In the 1990s the authorities never seemed to consider this question seriously. Their line of thinking was probably that it would be dangerous to escalate the 'whale war', considering that Norway's commercial catch was at least tacitly accepted by most actors internationally.

As the decade was coming to a close, the issue was being more seriously considered. Again the Ministry of Foreign Affairs was reluctant to resume export, fearing international repercussions, but the Prime Minister's Office was again more positive to the demands from the whalers. Ms. Brundtland was now out of office, but it was another Labor Government that made the decision to resume commercial export. The new Centrist Government, however, has upheld the Labor Government position, thereby confirming the consensus behind the more offensive Norwegian whaling position.[23]

The Parliament also played an active role in this process, first spearheaded by the Coastal Party, established in 1997, which was particularly concerned about living conditions and vocational opportunities in the coastal areas. Mr. Bastesen and the right wing Progressive Party pushed the issue in Parliament, but there were no objections from other parties. Overall, there was far less attention surrounding the opening up of export compared to the decision to resume commercial whaling – a reflection of the increasing marginalization of the whaling issue both domestically and internationally.

According to well placed sources, it was not only a calculation of political and economic costs and benefits and pressure from the target group that was behind this decision.[24] There was a legal aspect as well. Mr. Bastesen questioned the legality of the Norwegian decision in 1989 to stop the export of whale products in the first place, and he wanted to try the issue before the court. The authorities had gotten an expert opinion on this issue, and there was serious doubt that they would win this trial. Therefore, to some extent the authorities were 'forced' to make this decision. In other words, the question of principle and who was 'right' or 'wrong' in the issue of resuming export was far less prominent than it was with the issue of resuming commercial whaling.

This concludes the story about the Norwegian decision-making processes regarding the halt of commercial whaling, the halt of whale product export, as well as the subsequent decisions to resume both. To a large extent these processes have been driven by a small number of key persons, and their main strategy has been to create *domestic* consensus as a means to embark upon a more offensive whaling policy. Could more have been achieved if there had been a domestic consensus during the whole period? Yes, probably, in the sense that Norwegian whaling would probably not have been stopped at all. Although the domestic game has been crucial to understanding this process, the international dimension was also of considerable importance throughout, as the next section will demonstrate.

The Significance of the International Dimension

It has been noted that there are several international processes and institutions relevant for the whaling issue. The following section focuses only on the three most important institutions as seen from a Norwegian whaling perspective: the IWC, CITES and NAMMCO.[25]

The core regime (IWC): still protectionist, but gradually more 'balanced'? When the IWC started tightening regulations, it had an immediate effect on Norwegian whaling policy. From 1976-86 IWC regulations were transformed into domestic legislation, and for the first time Norway adopted a total quota for its small-scale whaling. Although Norway reluctantly accepted the strong *reduction* of quotas, it would not accept a complete *stop* in commercial whaling. Therefore, Norway objected to the 1982 moratorium and also used its right under the Convention to reserve itself against the decision in 1985 that the northeast Atlantic minke whale stock should be classified as a protected stock, meaning that it could not be harvested. According to the IWC moratorium decision, the Norwegian catch of minke whale should stop by 1986. Norway, however, declined to do so and adopted a *unilateral* quota of 400 animals based on population estimates from the IMR. Thus, Norway initially went along with the IWC majority for a while, but then refused to go further before harvest was temporarily stopped after the 1987 season.

As has been discussed above, subsequent direct IWC influence over Norwegian policy was reduced as whaling was resumed. Resolutions, without legal impact but potentially politically important, were routinely adopted at the yearly IWC meetings. First they were directed against Norwegian scientific whaling and later against commercial whaling. Although these had no direct impact on Norwegian policy, Norwegian delegates worked hard to prevent their passage, and vocally protested their content. When such resolutions were successfully blocked, it was seen as a small, but important victory for the Norwegian position. No doubt one reason Norway maintained a moderate catch in line with the very cautious

IWC management procedure was to weaken the basis for these critical resolutions. Still, the fact that a compact majority was against Norwegian whaling illustrated that, overall, the international dimension of the political strategy launched to defend and explain the decision to resume whaling was less successful than the domestic overtures.

As a point of departure, Norway has had modest success in affecting the goals and policies of IWC. The main reason is simple: The anti-whaling nations have used the *voting mechanism* very actively to outvote the minority on major as well as minor issues. This has been described as the 'tyranny of the majority', as no considerations are made for the affected minority (Friedheim, 2001). That is, instead of seeking *consensus*, the normal procedure and approach within international regimes, the IWC resembles a domestic parliament where the majority rules.

During the past decade, however, the balance between pro-whaling nations and anti-whaling nations has gradually shifted. In addition, there has been an increase in the number of nations that are not readily placed in either of these two categories. In 1982 when the moratorium was adopted there were only six nations voting against the moratorium – and most of these later withdrew their objection.[26] More recently, the votes have been divided much more evenly. This development is due in part to an active 'recruitment policy' on the part of the pro-whaling side, most notably Japan. That is, the strategy employed by anti-whaling states and green NGOs to 'pack' the Commission so as to pass the moratorium in the late 1970s and early 1980s has later been adopted by the other side. In addition, many of the previous anti-whaling nations have left the IWC. However, the anti-whaling movement has reacted to this development, and they have started to recruit new members. This means that the overall number of whaling nations is finally on the rise after some two decades of stable membership of about 40 nations.[27]

The upshot of this is that neither side is close to commanding the three-quarter majority that made the moratorium possible. On some issues there is now in fact close to a 50/50 vote, so the anti-whalers can no longer control the agenda as they used to. This was clearly demonstrated at the IWC meeting in 2001, when a majority of only one vote denied membership to Iceland. Iceland's application for membership was conditional on its having the right to reserve itself against the 1982 moratorium.[28] At the latest IWC meeting in Berlin (2003), however, the anti-whaling nations showed that they were still in command through the 'Berlin initiative' and the establishment of the Conservation Committee – much to the anger of the pro-whaling forces. They claim that the purpose of the committee is to change the very mandate of the IWC – into a pure conservationist body. Twenty-five members were in favor of the proposal, 20 were against and one member state abstained. The more radical forces once again claim that this should be the end of the IWC and that alternative management bodies need to be set up (High North Alliance, 2003). In my opinion, the likelihood of this happening is very small. The issue is simply not important enough for a significant number of important actors. That is, the IWC will probably continue to limp along the bumpy road of conflicts between conservation and utilization.

Irrespective of the continuing command the pro-whaling forces still have over the IWC, there is no doubt that Norway has played an important role in the development of a more balanced IWC. Around 1990 it seemed the battle over the future of the IWC had been won by the anti-whaling forces. These forces were completely in command of the discussion and seemed to have the moral upper hand. Although the whaling side is still in the minority, the picture has changed significantly over the last decade. The pro-whaling forces, with a strong developing country component, were much more on the offensive in terms of force of argument as well as self awareness. Norway unquestionably played a significant role in this turn of the tide. In line with its tradition, however, Norway has been a very 'decent' player. There has been no active recruitment through 'sweeteners', and Norway also voted in favor of granting quotas to the aboriginal peoples in the US and Russia at the 2002 IWC meeting because this was seen as the 'right' decision.[29] That is, Norway relies more on persuasion and knowledge than on tactics and threats. To substantiate these claims, it can be noted that the previous IWC Secretary for some 30 years claimed that in his opinion among the pro-whaling members Norway was always a well-prepared actor with a high level of expertise among its delegates, very rarely making the kinds of mistakes or blunders that most other (pro) whaling forces made once in a while.[30] In this sense, Norway has contributed to increasing the legitimacy of the pro-whaling side, thereby contributing to the gradual turn of the tide in the IWC. No doubt, some would claim that Norway in the IWC qualifies as an instrumental leader (Young, 1991). Due to the polarized nature of the issue, others would disagree strongly – maintaining that Norway could at best be labeled a leader for the pro-whaling forces, or more of a group leader than a true leader.

Linked regimes: NAMMCO and CITES, bargaining chips, and 'venue shopping'[31]

Due to its moderate success in changing the IWC from within, Norway has been active in using other fora to pursue its interests and increase goal attainment. Before turning to the two most important arenas in this regard, NAMMCO and CITES, first a few words about other relevant international institutions.

The Law of the Sea negotiations was a top priority in Norway during the 1970s, and Norway played a leading role during parts of the negotiations (Skodvin, 1992). However, the whaling issue received very little attention as other issues were considered far more important, and the IWC was the main forum for management of whales. Still, it is important that this major instrument governing all uses of the sea and its resources did not reinforce the anti-whaling stand. The right of the coastal states to decide whale management was underlined, and the negotiations also opened up for other management bodies than the IWC. While the Stockholm Conference was a landmark against commercial whaling, the Rio 1992 Summit did not reinforce the protectionist stance of the IWC majority as sustainable use, not protection, was the main take-away message from that Conference. At the 2002 Johannesburg Summit, the issue was not even on the agenda. It was discussed behind closed doors between a limited number of actors, illustrating the increasing marginalization of the issue.[32] As to GATT/WTO rules in relation to whaling and marine mammals, suffice it to note that legally the rules may protect Norwegian whaling interests, as they make enacting sanctions against whaling countries more difficult. The political implications, however, are more uncertain (McDorman, 1997). It is also important to note that all of these institutions are far more representative than the IWC because their memberships and levels of participation by far exceed that of the IWC. In general this has contributed to increasing the legitimacy of the Norwegian pro-whaling position somewhat. Having only some 45 members with voting rights, the IWC is probably the world's smallest global organization. The IWC, however, is a huge organization compared to NAMMCO.

NAMMCO was established in 1992 by Norway, Iceland, the Faroe Islands and Greenland.[33] That is, NAMCCO consists of only two sovereign countries. The two other members have some independence but are still linked to Denmark. Both are whaling nations, however, and rather distant from their southern 'motherland' in terms of whaling interests and whaling policy.[34] When Iceland left the IWC in 1991, it no doubt saw NAMMCO as a management alternative to the IWC. The more radical parts of the Norwegian fisheries sector shared this opinion. The Norwegian authorities, however, did not share this view, at least not after Norway started commercial whaling while still being an IWC member. That is, for the last decade Norway has to my knowledge never really considered leaving the IWC and has tended to use NAMMCO more as a 'bargaining chip'. If the IWC majority has been 'too unreasonable', Norway has threatened to leave and rely on NAMMCO as a management alternative.

As it became clear that NAMMCO was no real alternative to the IWC, Iceland decided to re-enter the IWC. Originally, there was hope that NAMMCO would attract more members and thereby provide a broader political base and represent a more realistic alternative to IWC. Both Canada and Russia have been observers, and there have been rumors that they would enter, but this has never materialized. NAMMCO is therefore a true 'mini-regime' that does not pose any real threat to an IWC with growing membership. The fact that not only Norway but also Iceland will conduct commercial whaling as members of the IWC will reduce the credibility and value of NAMMCO both as a bargaining chip and as a potential management alternative. Its prime significance at present is knowledge building and its function as a forum for deliberation between pro-whaling forces in a peaceful and friendly atmosphere, in sharp contrast to the hostile IWC atmosphere.[35]

CITES was signed in 1973. Its main objective is to prevent trade in endangered animals. CITES introduced a listing system, depending upon the protection the various species needed. Annex 1 lists species that are threatened with extinction, while Annex II lists those that are not necessarily threatened with extinction but where regulations are needed. Although CITES is very important for the whaling issue, the issue of whaling is a minor one in CITES. To indicate the proportions, more than 30,000 species are listed on the CITES agenda. However, most attention – by Western countries and Western NGOs – has traditionally been paid to the so-called charismatic mega-fauna, including mostly exotic tropical wild animals, but also marine mammals.

Turning to the link to whales, a key question was whether CITES should more or less automatically follow IWC in its listing policy, or whether more independent evaluations should be done. Initially there were some discrepancies between the two, so the CITES Secretariat was instructed to consult with the IWC concerning future amendments to avoid conflict of regulation. So, in line with the 1982 moratorium CITES included all IWC Protected Stocks in Appendix 1 at the 1983 CITES meeting. There are rights of reservations, and Norway as well as Japan (and others) have made reservations against the whaling listing and can therefore *legally* trade between themselves; *politically*, however, this is another matter. During the 1980s CITES therefore effectively reinforced the anti-whaling message.

In the 1990s this changed somewhat. First, it used to be virtually impossible to obtain transfer from Appendix I to Appendix II, but this has been changed since 1992. Second, the large scientific whaling programs launched by Norway (and Japan) gave more substance to the demand for changing the listing of the minke whales to Appendix II. A third trend is equally important for understanding the gradually changing sentiment: To cut a long and complex story short, many developing countries as well as the whaling countries came to realize that they were in the same boat. They had both had policies they did not believe in forced upon them, policies that were not based on science but rather on values they did not agree with. A typical case in point was Norway. Like most 'green' Western countries, it had traditionally supported the unconditional protection of the elephant. However, when it was realized that this position was no more scientifically based than the view that all whales were depleted, Norway changed its standpoint on the issue.

While the IWC has proven to be quite static and entrenched in deep conflicts, CITES has been more dynamic with its 160 member countries. Thus, while the initial strategy of the pro-whaling forces was to keep CITES away from IWC turf, in the 1990s they came to see an opportunity for using CITES as an instrument to further their interests. Although this strategy has not been an outright success, it has at least been more successful than their efforts in the IWC. At the 9th Conference of the Parties (CoP 9) in 1994 the proposal to down-list the Northeast Atlantic minke whale was defeated by a large margin (48 to 16 with 49 abstentions). In contrast, at CoP 10 three years later, the Norwegian proposal was accepted with a wide margin, but the required two-thirds majority to down-list this species was not within reach. Still, considering this positive development, Norwegian lobbying was intense prior to CoP 10 in 2000. However, at this CoP the support for the Norwegian proposal was reduced and it was accepted with a margin of only 1 vote. The requited two-third majority had once more become a distant dream.

Nevertheless, Norway could claim that a *majority* was behind the Norwegian position, and this was a much more representative majority than the anti-whalers could muster in the IWC, considering its limited size. This gave Norway some political backing and legitimacy, although not to the extent that it had hoped and lobbied for. In that sense, CITES has played a role in the decision to resume export, but other factors have also been important. At the most recent CITES meeting in Chile in 2002, Norway did not present any specific proposals regarding down-listing. This was linked to the fact that as export had opened up, there was really no point in asking once more for a CITES opinion, especially as there was no chance of getting a two-thirds majority.[36]

Overall, CITES has had a significant impact on Norwegian whaling policy, first by contributing to stopping export and later by enabling the reopening of export. CITES has also been an important learning ground for building new alliances, most notably with developing countries.

A final note on the role played by Norway and the priority given to the CITES negotiations.[37] Overall, Norway has probably been instrumental in the gradual turn of the tide in CITES as well – together with other 'sustainable use countries'. As in the IWC they were outvoted in the 1980s, but gradually got their act and strategy together in the 1990s. More specifically, the whaling issue has been *the* single most important issue in CITES directly affecting Norwegian interests. Overall, trade in endangered species is not an issue of much relevance to Norway. However, the more *principle* matter of sustainable use has been important, not least linked to adjoining Conventions like the Convention on Biological Diversity (CBD). This is probably one of the reasons the Norwegian head of delegation comes from the environment sector and not the Ministry of Foreign Affairs or fisheries, although the latter issue is becoming increasingly important in CITES. This may also be a clever tactical move to tone down the Norwegian emphasis on whaling in CITES somewhat, providing Norway with a somewhat softer and 'greener' image than is the case in the IWC.[38]

Alternative Explanatory Perspective: Problem Structure and the Key Role of the US

As a point of departure the whaling issue is presently a very benign issue, as the economic stakes involved are extremely modest and very few are directly affected by whaling regulations. This is, however, one side of the coin only. In political terms the whaling issue is extremely malign in the sense that polarization is so high and the value dimension so strong. For parts of the public as well as the green NGO community in most of the anti-whaling countries, the whale is a mythical animal, a key member of the charismatic megafauna family of exclusive animals (Skodvin and Andresen, 2003). One might watch the whales, adopt whales or listen to their songs. It goes without saying that within such a perspective one should not kill whales. It has been argued that under such circumstances negotiations are extremely difficult and there are slim chances of introducing a 'rational' element like science (Andresen, 2000, Andresen, 2001b). No doubt this rather unique and malign problem structure is very important in explaining the difficulty Norway has had in explaining and defending Norwegian whaling policy within the IWC.

Nowhere has this perception on the mythical role of the whales gained stronger ground than in the US.[39] Therefore, Norwegian whaling policy cannot be understood properly unless the role of the US is briefly discussed.

Neither the IWC nor any other international institution has been the most important for Norwegian whaling policy. The most important actor has been the US due to its 'policeman role' in the issue, enacted through its will and ability to use political, legal and economic measures to stop commercial whaling. US threats were the main reason Norway stopped commercial whaling in 1987, not 'hate resolutions' from the IWC or any other fora. In short, the US paid particular attention to Norway – and vice versa. Bilateral consultations behind closed doors, in a very hostile atmosphere, were frequently conducted in the late 1980s and early 1990s, and Norway was certified for sanctions four times in this period by the US. Finally a tacit agreement or understanding was reached at the very highest political level, between Prime Minister Brundtland and Vice President Al Gore. The understanding seems to have been that Norway would keep its catch at a low level and retain its membership in the IWC. Still, it was by no means self-evident that the US would refrain from reactions when Norway resumed commercial whaling. The decision of whether or not the US would attempt to levy sanctions was not resolved until it had gone all the way up to the President.[40]

Thereafter, this bilateral dimension has become less significant, as the whaling issue has moved into more smooth political waters. The two parties are on less than friendly terms in the IWC and other relevant fora, but this is mostly seen as 'routine policy': the two have agreed to disagree. The US did not threaten Norway with sanctions when export was resumed. In short, US whaling policy was maybe the most important reason why Norway stopped whaling. Over time, however, it appears that Norway has also been able to influence and modify the US position somewhat. Presently the US is no longer so important for Norwegian whaling policy, but this may easily change if Norway escalates the whale war further, or the issue regains prominence on the international agenda.

Conclusion

The whaling issue illustrates the limited status of rationality in international politics – when values and principles are at stake. From a rational cost-benefit perspective, economically as well as politically, it clearly would have made most sense for Norway to stop commercial whaling. Economically, whaling meant next to nothing for a wealthy country like Norway, and the potential economic and political costs of going against all major allies, as well as the green movement, were considerable. Moreover, Norway's green image could have been severely damaged. Nevertheless, after a thorough decision-making process, Norway declared that its official goal was to resume whaling after an interlude of five years.

A 'medium score' has been assigned to Norwegian goal attainment. Attainment is fairly high in the sense that whaling (and export) has been resumed, and the political and economic costs have been very small. In terms of influencing the core regime, however, success has been somewhat lower.

In the mid 1980s the whaling issue was about to become a real bone of contention on the domestic scene. After having been the exclusive domain of the fisheries sector, international polarization triggered strongly increased domestic participation, with most of these new actors quite negative to continued whaling. A high-level scientific-political strategy, orchestrated by the Prime Minister herself, however, managed to rally domestic support to resume whaling. No doubt the clear scientific message was important here, as well as the strong international condemnation which was felt to be unfair and therefore bolstered support behind the

Norwegian position. The issue had become one of principle – of rational use of natural resources as well as national sovereignty – and Norway was not ready to give in. The fact that no significant negative consequences emerged made it easy to maintain the domestic consensus.

Internationally, Norway has worked relentlessly and also consistently in various fora to defend the Norwegian position. Norway has had a high profile on this issue, and has used much political energy to promote its interests. As a small country, Norway has not been able to turn the tide in the IWC, but through active 'venue shopping' (through CITES), bargaining chips (NAMMCO) and a very high profile in the IWC, some scientific and political results have been achieved. High consistency and energy has been facilitated by the fact that although the relevant fora have been many, the key Norwegian players have been few. When it became clear that no important sectors of Norway were hurt by the more aggressive policy, the issue has been left essentially to a very limited number of people in the Ministry of Foreign Affairs and the Ministry of Fisheries, with close links to the Whalers Union and Fishermen's Organization. The Prime Minister's Office only enters the scene when there are questions of principle at stake, and the Ministry of the Environment, potentially less enthusiastic about the Norwegian position, has generally been marginalized. A few key people stand out as particularly important: Ms Brundtland for making the decision to resume whaling, as few Norwegian Prime Ministers would have done; Professor Walløe for his consistent engagement and expertise; and finally Mr Bastesen for his ability to keep the issue alive in the Parliament as well as among the public.

It is not self-evident, however, that the present rather idyllic domestic picture will last. There are some dormant, but quite important, forces that are less happy with the present Norwegian policy, both in the Ministry of the Environment as well as in the Ministry of Foreign Affairs. Partly they dislike the Norwegian aggressive policy; partly they think too much political energy has already been spent on this marginal issue. Other much more important issues get less political attention and resources, they argue. These forces may be triggered if external opposition grows once more – or if the whaling issue is moved closer to other environmental concerns by being considered to be more closely linked to biodiversity than fisheries.

Notes

1 This section is based primarily on Tønnessen and Johnsen (1982) and Birnie (1985).

2 The exception was the Japanese whalers, well known for their effective utilization of the whales, including use of the whale meat.

3 Germany quit whaling after World War II.

4 Some nations abstained, but Norway voted *in favor* of the resolution, probably due to the fact that Norway was represented by the Ministry of the Environment.

5 As noted, Germany never started whaling again, but Japan was allowed by the US to start hunting as hunger and poverty were widespread in post-war Japan, and it rejoined the IWC in 1951.

6 Although the US was no longer a major whaling nation at the time, it had some scientific expertise and the International Whaling Convention to a large extent was based on a draft elaborated by the US biologist Kellogg O'Brian.

7 For an elaboration of the basis for the differentiation between phases, see Andresen (2001b).

8 The moratorium was to be implemented by 1986 at the latest, and its effect should be evaluated by 1990 at the latest in light of a comprehensive assessment, but this assessment has never been carried out due to a lack of will on behalf of the IWC majority.

9 This does not necessarily mean that the IWC is a 'high-effectiveness' regime. This depends upon the measuring rod applied (Andresen, 2001b).

10 Personal communication, Ministry of Foreign Affairs, 1990.

11 Previously there had been suspicion of illegal trade with minke whale products, including instances where Norwegian whalers had been involved.

12 Based on access to internal Norwegian documents.

13 This section relies heavily on Andresen (1998).

14 Personal communication with previous Norwegian IWC Commissioner, 2002.

15 Their views were important inputs in the process of setting up NAMMCO.

16 The output of this process was *Our Common Future* (The Brundtland Report) (World Commission on Sustainable Development, 1987), probably the most important international environmental document of the 1980s.

17 No doubt he was brought in due to his scientific expertise and position, but he was also a personal friend of Ms. Brundtland.

18 He also brought in other new scientists, experts on whale population dynamics, not previously involved in this type of research in Norway. This type of expertise was crucial to balance similar expertise of scientists opposed to commercial whaling (Schweder, 2000).

19 Norway's relationship with the US will be described in greater detail in the section on problem type.

20 Based on information from key players, the EU process was not decisive for the decision made. Moreover, it had little or no impact on the coastal communities' attitudes towards the EU. They voted against joining the union, and so did the majority of those casting their votes.

21 This is quite a common reaction to external threats when the issue is coined in terms of important principles and national sovereignty (Knorr, 1978).

22 Usually the Commissioner comes from the Ministry of Foreign Affairs, keeps this position for a few years, and often has little or no knowledge about the issue when he or she starts. In contrast, professor Walløe has been on the delegation every year since 1987.

23 When the Norwegian Prime Minister visited Japan in 2002, the issue was discussed at the highest political level.

24 Interview in the Ministry of Foreign Affairs, summer 2002.

25 If and when Norwegian EU membership is once more on the agenda, EU rules and regulation will become crucial for continued Norwegian whaling policy. As the EU membership issue is currently not on Norway's agenda, it will not be included here.

26 The only parties that did not were Norway and Russia.

27 There are now (fall 2003) 50 members, but all these do not have voting rights due to failure to pay membership fees.

28 Iceland had originally accepted the moratorium, but left the IWC in 1991 due to frustration with the development of the IWC. The renewed application with its reservation proved to be a thorny legal question. In the end, however, the political compromise was that Iceland was allowed to enter with the objection, but was required to promise to not start commercial whaling until 2006. Scientific whaling, however, started this summer.

29 A rare alliance between countries fundamentally opposed to all kinds of whaling (like Australia, the UK and New Zealand) and quite a few of the pro-whaling forces voted against giving these quotas to the USA and Russia. This had never happened before. However, at an extraordinary IWC meeting later that year this decision was reversed.

30 Personal communication with Dr. Ray Gambell, Cambridge, summer 1998.

31 For an elaboration of the significance of other 'linked' regimes relevant to whaling, see Andresen (1997) and Andresen (1999).

32 Personal communication with WWF representative following this issue at the Johannesburg Summit.

33 NAMMCO manages seals, provides inspection of whaling done by the NAMMCO members, and conducts research.

34 The aboriginal catch of Greenland is regulated by the IWC, while the catch of pilot whales in the Faroe Islands is outside the scope of IWC regulations.

35 Japan and some of the pro-whaling Caribbean countries also often participate as observers.

36 Japan presented various down-listing proposals on whaling but they were all defeated.

37 This is based primarily on impressions from the last CITES meeting in 2002 where I was an (observing) member of the Norwegian delegation.

38 Still, most specific attention is on whaling and fishing in the delegation. The more lofty ecological principles tend to get less attention. The delegation is quite small, compared, for example, with the delegation to the IWC.

39 Few perceive the US as a whaling nation, as it is so vehemently against commercial whaling. The US is, however, a significant whaling nation due to its aboriginal catch, permitted by the whaling convention. This has made many accuse the US of a double standard in its whaling policy (Andresen, 2001b).

40 Interview with a US official that took part in this process, spring 1998.

References

Anderson, R. (1987), The State of the Northeast Atlantic Minke Whale Stock, Økoforsk, Norway.

Andresen, S. (1997), 'NAMMCO, IWC and the Nordic Countries', in Whaling in the North Atlantic, G. Petursdottir (ed.), Fisheries Research Institute, University of Iceland, pp. 75-89.

Andresen, S. (1998), 'The Making and Implementation of Whaling Policies: Does Participation Make a Difference?', in D.G. Victor, K. Raustiala and B. Skolnikoff (eds), The Implementation and Effectiveness of International Environmental Commitments, MIT Press, Cambridge MA and London UK, pp. 431-75.

Andresen, S. (1999), 'The International Whaling Regime: Order at the Turn of the Century?', in D. Vidas and W. Østreng (eds), Order for the Oceans at the Turn of the Century, Kluwer Law International, The Hague, London and Boston, pp. 215-28.

Andresen, S. (2000), 'The Whaling Regime', in S. Andresen, T. Skodvin, A. Underdal and J. Wettestad (eds), Science and Politics in International Environmental Regimes, Manchester University Press, Manchester and New York, pp. 35-70.

Andresen, S. (2001a), 'The International Whaling Regime: "Good" Institutions but "Bad" Politics?', in R. Friedheim (ed.), Towards a Sustainable Whaling Regime, University of Washington Press, Seattle and London, pp. 235-69.

Andresen, S. (2001b), 'The International Whaling Commission: More Failure than Success?', in E.L. Miles, A. Underdal, S. Andresen, J. Wettestad, J.B. Skjærseth and E.M. Carlin, Environmental Regime Effectiveness: Confronting Theory with Evidence, MIT Press, Cambridge MA and London UK, pp. 379-405.

Birnie, P. (1985), International Regulation of Whaling, Volume 1 and 2, Oceana Publications, Inc., New York, London and Rome.

Friedheim, R. (2001), 'Negotiating in the IWC Environment', in R. Friedheim, Towards a Sustainable Whaling Regime, University of Washington Press, Seattle and London, 200-35.

High North Alliance (19 June, 2003), 'End in sight for the IWC?', retreived on 20 January 2004 from www.highnorth.no/news/nedit.asp?which=309.

Knorr, K. (1978), 'International Economic Leverages and Its Uses', in K. Knorr and F. Trager, Economic Issues and National Security, University Press of Kansas, Laurence KA, pp. 99-125.

McDorman, T. (1997), 'Iceland, Whaling and the US. Pelly Amendment: The International Trade Law', Context, Nordic Journal of International Law, vol. 66, pp. 453-74.

Ministry of the Environment (1989), Environment and Development. Programme for Norway's Follow-up of the Report of the World Commission on Environment and Development (Miljø og Utvikling. Norges oppfølging av Verdenskommisjonens rapport), Report No. 46 to the Storting (1988-89), Ministry of the Environment, Oslo.

Norwegian Parliament Debates (Forhandlinger i Stortinget) (1993), nr. 255, 18 May, pp. 3853-67, The Norwegian Parliament, Forvaltningstjenestene, Oslo.

Schweder, T. (2000), 'Distortions of Uncertainty in Science: Antarctic Fin Whales in the 1950s', Journal of International Wildlife Law and Policy, vol. 3, nr. 1, pp. 73-92.

Skodvin, T. (1992), 'Structure and Agent in Institutional Bargaining', Cooperation and Conflict, vol. 27, nr. 2, pp. 163-89.

Skodvin, T. and Andersen, S. (2003), 'Non-State Influence in the International Whaling Commission', Global Environmental Politics, vol. 3, no. 4.

Tønnessen J.N. and Johnsen, A.O. (1982), The History of Modern Whaling, C. Hurst & Company, London.

World Commission on Environment and Development (Verdenskommisjonen for miljø og utvikling) (1987), Our Common Future (Vår felles framtid), Tiden Norsk Forlag, Oslo.

Young, O. (1991), 'Political Leadership and Regime Formation: On the Development of Institutions in International Society', International Organizations, vol. 45, nr. 3, pp. 281-308.

Chapter 4

Ozone: A Success Story on all Fronts?

Tom Næss

Introduction

Ozone-depleting substances (ODS) in the Earth's atmosphere were first discovered and brought to the attention of the public by two American scientists in 1974 (Molina and Rowlands, 1974). While the ozone case received mostly lip service during the 1970s, new scientific discoveries marked the early and mid 1980s, and international negotiations resulted in the adoption of the Montreal Protocol in 1987 on reduced production and consumption of major ODS (chlorofluorocarbons (CFCs) and halons).[1] The Montreal Protocol has produced more and more stringent regulations on ODS and has also proven to be flexible in providing governments with effective instruments to reduce emissions of ODS on a global scale (Oberthur, 1999).

Since the early 1980s, Norway has played an important role in promoting international efforts to reduce potential damage to the ozone layer. With other partners in the Toronto Group,[2] Norway has pushed consistently for global reductions in the emissions of ODS. Judging by a global 85 per cent reduction in the production and consumption of these substances, the Toronto Group has succeeded, and the ozone layer has stopped thinning and will recover partially in 2050 (WMO, 1999; Oberthür, 2001). At the international level, Norwegian ministers and officials have worked hard to establish a regulatory framework for international action through the Vienna Convention for the Protection of the Ozone Layer of 1985 and the Montreal Protocol on Substances Depleting the Ozone Layer of 1987. Norway's foreign environmental policy has clearly been to strengthen the international regulations on the emissions of major ODS, and through the efforts of the Toronto Group, Norway has succeeded in attaining that goal.

At the domestic level, the Norwegian government has adopted various means of phasing out ODS which have proven to be effective. The success in tackling the 'ozone problem' from both a global and a domestic perspective corresponds well with Norwegian environmental goals, suggesting that Norwegian goal attainment within this issue area is high. At the national level, the Norwegian Government's objectives are spelled out in the 1988-89 White Paper (Ministry of the Environment, 1989) on Norway's follow-up to the World Commission on Environment and Development (WCED): a 50 per cent reduction, relative to 1986 levels, of CFC emissions by 1991; and by 1 January 1995, CFC emissions reduced by 90 per cent and a complete phase-out of halons. These ambitious objectives were easily achieved, and today domestic consumption of ODS has been reduced by 98 per cent (PCA, 2001a).

Compared with other environmental problems such as climate change and air pollution, with respect to the 'ozone problem', effective results were achieved both in Norway and in many other countries in a short period of time. This chapter sets out to explain how and why Norway managed to solve this environmental problem so effectively in just a few years. It argues that a lack of 'crossfire', or rather a sort of 'positive crossfire' where other linked regimes, the core regime (the Montreal Protocol) and national interests have influenced Norwegian ozone policy positively, lies at the heart of the explanation.

In the following section, a brief overview of Norwegian ozone policy, goals and achievements will be given. In the first part of the next section I discuss Norway's role in the formation of the core regime, the Montreal Protocol, and this regime's role in promoting effective phase-out efforts at nation-state level, and other linked regimes influence on the formation of a Norwegian ozone policy. Then I will explain why ozone policy is a success story in Norway. Due to a set of factors, beneficial circumstances and government decisions Norway has rid itself of the ozone problem more or less entirely. Towards the end of the study, I will sum up what we believe are the main factors behind Norway's goal achievement within this issue area. Neither linked regimes, nor national interests have obstructed Norway from pursuing ambitious political phase-out goals.

Norway's Ozone Policy: Ambitious Goals and Successful Achievements

Norway's goals for protecting the ozone layer have been ambitious from the start. This level of ambition makes the level of achievement all the more remarkable. This section provides an overview of Norway's goals at both the international and national level, and the degree to which they have been attained.

Goals

A long-term, official Norwegian goal has since the early 1980s been to reduce the threat to the ozone layer by decreasing emissions of ODS to the atmosphere (Vaggen Malvik, 1998; PCA, 2001a). As a result, Norway was among the first countries to implement regulations on the consumption of ODS. Already in 1981, the use of CFCs in spray cans was prohibited, but the first comprehensive plan for phasing out existing use of ODS came in an Action Plan adopted in 1988, where the government proposed phasing out the use of CFCs by 50 per cent in 1991, and by 90-100 per cent in 1995. The Norwegian government decided to threaten the industry with a tax on CFCs and halons (40 Norwegian kroner per kilo). The tax was to be implemented by 1991 if industry did not reduce imports in accordance with set objectives. Due to a willingness on the part of industry to reduce its consumption of CFCs and halons, and a temporary slump in the Norwegian economy which led to reduced demand for ODS in industry and construction, the government never implemented the tax. Instead of implementing the tax, a long-term policy was implemented in 1991 to phase-out ODS in accordance with agreements made under the Montreal Protocol. This policy contained a plan to limit the import, export and consumption of CFC and halons in Norway (Ministry of the Environment, 1995).

Norway's adoption of the Action Plan and the act of 1991 must be seen in light of international negotiations on the Montreal Protocol in 1987, and the standing of the Toronto Group in these negotiations. The Toronto Group pushed for strict regulations, but met opposition from a number of other states (Benedick, 1991). The initial agreements reached in September 1987 reflected a breakthrough in the discussions. It was a compromise between forces wanting to protect the ozone layer by regulating emissions, and several states, backed by larger ozone-producing companies, that resisted regulation because of the scientific uncertainty.

Through the 1980s and 1990s, Norway's ozone policy, at both the domestic and the international levels, has evolved in line with the international development on the question of regulating ODS. The Montreal Protocol has been the most important process in that regard. The negotiations leading to the adoption of the Vienna Convention, the Montreal Protocol and consecutive protocol amendments have spurred domestic action in a number of countries to phase-out ODS, and Norwegian laws and regulations drafted in the late 1980s were clearly inspired by the outcome of negotiations. Norway actively pursued ozone policy up until the early 1990s, but beyond 1990 changes to the Norwegian ozone policy have been implemented as a result of international events. Norwegian ozone regulations were reformulated in 1991, 1995, 1996, 1997 and 2000, usually as a response to amendments and decisions made at Meetings of Parties under the Montreal Protocol (Vaggen Malvik, 1998; PCA, 2001a).

In the second half of the 1990s, Norway's adherence to the European single market through the European Economic Area (EEA) agreement in 1994 has affected domestic ozone policies (Dahl, 1999). EU member states formulated a comprehensive legislative act, Regulation 3093/94/EC, which directly affected Norway by encompassing more substances than the first Norwegian regulation covered. The Norwegian regulation of 21 January 1991 covered CFCs and halons only, whereas the EC Regulation 3093/94/EC covered these substances as well as several others including methyl bromide and hydrochlorofluorocarbons (HCFCs). In other words, even though it might seem that Norwegian and EC regulations contain parallel objectives, the substances covered and the contents of the regulations differ. (See Table 4.1) Thus, on one hand, Norwegian ozone policy became stricter in terms of including more substances and went further in controlling sales and import. On the other, EC regulations have gradually improved over time and expanded to new substances and new areas, thus gradually expanding and tightening Norwegian regulations on some issues such as HCFCs. Thus, Norwegian goal attainment has become stricter and more comprehensive over time, partly as a result of decisions made unilaterally by the Norwegian government, and partly as a result of international factors: the development of regulations under the Montreal Protocol and within the EU.[3]

Table 4.1 Phase-out schedules in Norway, under the Montreal Protocol and in the European Union

	CFCs	Halons	HCFCs*	Methyl Bromide	Carbon Tertrachloride and Trichlorethan
Norway	100% reduction by 1.1.1995	100% reduction by 1.1.1994	100% reduction in consumption by 1.1.2015	100% reduction by 1.1.2005	100% reduction by 1.1.1996
Montreal Protocol	100% reduction by 1.1.1996	100% reduction by 1.1.1994	100% reduction in consumption by 2030	100% reduction by 1.1.2005	100% reduction by 1.1.1996
EU	100% reduction by 1.1.1995	100% reduction by 1.1.1994	100% reduction in consumption by 1.1.2010	100% reduction by 1.1.2005	100% reduction by 1.1.1996

* The consumption of HCFCs will be banned from 1.1.2010 in Norway and the EU, but production can continue for export to developing countries and for essential uses until 1.1.2025.

Sources: Ministry of the Environment, 1995; Rowlands, 1998; EC Regulation 2037/2000/EEC and PCA, 2001a.

Achievements

At the international level, Norway and other members of the Toronto group stressed the need for an immediate ban on aerosols as early as 1985. Even though this proposition met strong resistance from the EC at the time, by 1990 this first proposal had been adopted by all major parties to the Montreal Protocol. Thus what seemed impossible to accomplish in 1985 was adopted in 1990 by all parties to the Protocol in London, where the first Amendment to the Protocol was negotiated. In that sense, Norway directly influenced regulative output at the international level through Montreal Protocol negotiations (Benedick, 1991; Vaggen Malvik, 1998; Bakken, 2001).

At the domestic level, implementing the Montreal Protocol through regulation, financial incentives, information dissemination and voluntary agreements with industry has been an unqualified success (PCA, 2001a). Further discussion on the formulation and implementation of the Norwegian ozone policy follows below. Norway has gone beyond its international commitments under the Montreal Protocol, and has not confronted any major problems in complying with the objectives set out in domestic regulations, EC regulations or international regulations. Norway has a high score in terms of goal attainment, which can be seen in the following figure.

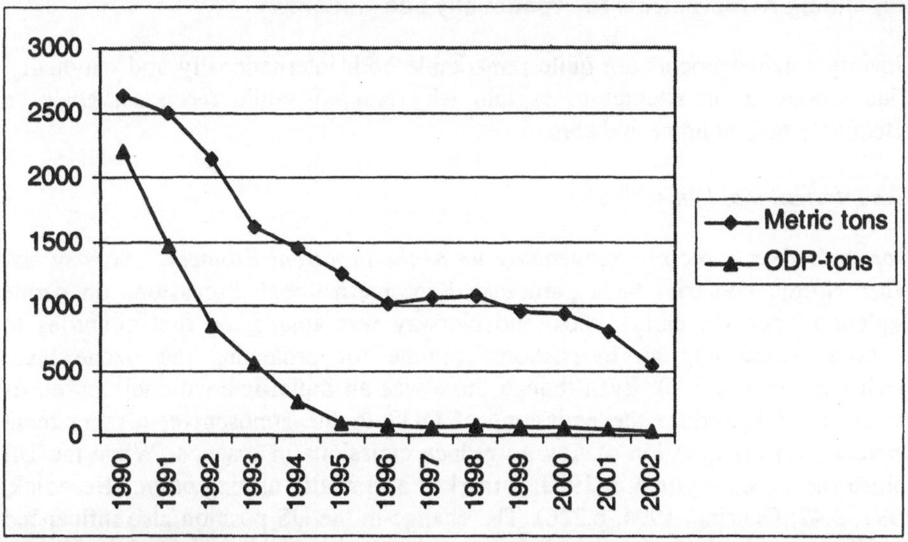

Source: PCA (2001a)

Figure 4.1 Norwegian imports of ozone-depleting substances 1990-2000

In 2000, Norwegian consumption of ODS was reduced by 98 per cent relative to 1986 levels. The remaining use of ODS is mainly related to essential use, quarantine and pre-shipment applications in accordance with Article 2 of the Montreal Protocol. In Norway, essential use is restricted to small amounts of CFCs, tetrachloride and carbon tetrachloride in laboratories, and small amounts of halons in fire extinguishing equipment aboard civil and military ships and aircraft. Quarantine and pre-shipment application comprises small amounts of methyl bromide for pest control in mills, ships and aircraft (PCA, 2001a). These amounts do not constitute any significance whatsoever on a global scale. However, in Norway and elsewhere in the Western world, there is a growing problem associated with the use of substitute fluorinated gases, mainly HCFCs and HFCs, as both are greenhouse gases. Norwegian imports of HCFCs, a substitute for CFCs, amounted to 936 tons in 2000, but decreased by 35 per cent from 2001 to 2002, which means the total import in 2002 totalled 540 tons. To reduce imports completely by 2010 will be a costly affair, according to the Norwegian Pollution Control Authority (PCA) (*Statens forurensningstilsyn*).[4] Nevertheless, Norway is obliged to phase out the substance before 2010 as this is the final phase-out date according to the newest European Union Regulation, 2037/00/EC. The situation on HFCs is yet another problem; however, it is more a problem for the climate than the ozone layer. The issue of HFCs will be further discussed under the section on interlinkages between the Kyoto and Montreal Protocols.

Explaining Achievements Internationally and Nationally

Norway's achievements are quite remarkable both internationally and nationally. This section is an attempt to explain why Norway could reach set goals so effectively both at home and abroad.

The International Dimension

Norway's role in the process leading up to the Montreal Protocol. Norway and other Nordic countries have participated in international discussions on ozone depletion since the early 1980s, and Norway was among the first countries to propose establishing an international regime for protecting the ozone layer (Gehring, 1994, p.200). Even though there was an initial international consensus on the need to reduce the emissions of ODS to the atmosphere, disagreement prevailed on the question of how to reduce emissions in practice. When the US joined the Toronto group in 1983, it marked a dramatic turning point. (Benedick, 1991, p.42; Gehring, 1994, p.226). The change in the US position also influenced the EC opposition to a world-wide ban, and led to the establishment of the Vienna Convention in 1985 and the adoption of the Montreal Protocol in 1987.

What limited Norway's capacity to achieve its goals in the first phase was the opposition of a majority of states, supported by the largest chemical companies in the world, to phasing out ODS. Due to the scientific uncertainty and the economic interests of the EC, which produced 38 per cent of all ODS consumed globally, the ozone issue was somewhat malign in this first period. Most larger European chemical companies such as ICI, Hoechst, Solvay, Elf Atochem and Ausimont were against phasing out ODS because they were large producers of ODS. But as the scientific knowledge became more clear-cut and the economic viability of substitute chemicals was confirmed, a growing number of states changed their views. When the US made its U-turn, US companies were forced to follow, and subsequently European companies were forced to follow as EC member states began to support the US position (Benedick, 1991). Consequently, the political proposals from Norway and the other Nordic states were taken up for discussion, and the main ideas behind the proposals were realized through the Montreal Protocol in 1987. Norway's role in the negotiations was linked in particular to participation in a smaller group consisting of heads of delegations and UNEP Executive Director Dr Mostafa Tolba that met for the first time in Geneva in April 1987. Dr Tolba's personal engagement in the meetings and the informality surrounding them proved instrumental in bringing negotiations on track. Apart from Norway, heads of delegations from Canada, Japan, New Zealand, the Soviet Union, the US, and the EC participated in the group. Mr Per M. Bakken, the Norwegian head of delegation, took part in the meetings with colleagues from the US, Canada, and New Zealand, and this group persuaded the EC, Japan and the Soviet Union to realise that they could not run away from emissions reductions. Eventually, this group proved instrumental in establishing a negotiation text that all parties could relate to, which ultimately led to the coming into force of the Montreal Protocol (Bakken, 2001). In other words, the political climate was changing internationally, and Norway was among the states taking advantage of the situation, coming forward with concrete proposals to reduce the emissions of ODS to the atmosphere. Thus, the openness of the international negotiations combined with Norway's skilled negotiators, who were able to formulate negotiable texts, facilitated Norwegian goal attainment and paved the way for an agreement in line with Norwegian interests.

Involvement of actors other than the Ministry of the Environment and the Norwegian Pollution Control Authority has been very restricted under the negotiations of the Vienna Convention and the Montreal Protocol. The Ministry of the Environment has been responsible for framing positions under international negotiations, and for Norwegian representation, whereas implementation, enforcement and the formulation of national regulations have been delegated to the Pollution Control Authority. Other ministries have participated to a limited extent in negotiations and the implementation process. The Ministry of Finance has taken part in international negotiations where issues of a financial character have been discussed, for example, issues related to yearly replenishment of the Multilateral Fund under the Montreal Protocol. The Ministry of Foreign Affairs, which normally is present where vital interests are at stake, has not been particularly involved in the negotiations, mainly because the negotiations have resulted in support of Norwegian positions. And when vital Norwegian interests have been met with support at the international level, there has been no need for the Ministry of Foreign Affairs to get involved (Bakken, 2001).

The Montreal Protocol – an effective international agreement. The core regime addressing ozone depletion consists of a framework convention addressing the issue in general terms (the 1985 Vienna Convention for the Protection of the Ozone Layer), and a protocol assigning practical, binding measures aimed at reducing the potential damage to the ozone layer globally (the 1987 Montreal Protocol on Ozone Depleting Substances). Looking into the substance of the agreement, the Montreal Protocol shares many similarities with other international treaties: an organisational apparatus (the secretariat), annual meetings of the parties (MoPs under the Montreal Protocol, CoPs under the Vienna Convention), technical assessment panels which meet frequently to discuss technical issues related to ozone depletion, and a set of rules on substance and procedure worked out over the years (agreements on various subjects reached by the parties). There is nothing special about these features; they are common to most international treaties. There are, however, three features that stand out as deserving comment.

First, the Montreal Protocol is among the first treaties to integrate science and policy by establishing panels and groups for scientific cooperation among parties. Thus, decisions on phase-out schedules and the control of an expanding group of ODS were all based on scientific results that the parties trusted. It included common but differentiated responsibilities, and it created new, innovative institutions such as the Implementation Committee and the Multilateral Fund. A website from the Canadian Global Change Program states:

> The innovative approach adopted by the Montreal Protocol in the field of rule-making and rule-implementation constitutes a major contribution to the development of international environmental law. Thus, ten years later, the pioneers of 1987 have bequeathed the international community a unique international instrument whose effectiveness continuously improves as scientific knowledge mandates and the political context allows. (CGCP, 2002)

The mechanisms invented under the Montreal Protocol were innovative and have proven to be effective in securing compliance. The mechanisms and panels devoted to scientific cooperation, economic aid, and non-compliance have made the Montreal Protocol one of the most effective environmental regimes (Porter and Brown, 2000, p.87). All Parties to the Montreal Protocol, including Norway, are obliged to report their production, import and export of ODS (Article 7). The data reported in accordance with Article 7 has enabled the secretariat to report cases of non-compliance to the Implementation Committee. The Committee is elected among the parties, and has a mandate which is to either assist the non-complying parties technically, economically, or otherwise, to issue cautions, or to suspend the party from its rights and privileges under the Protocol.[5] The Implementation Committee, the Multilateral Fund, the financing mechanisms of the Montreal Protocol, and the Technology and Economic Assessment Panel (TEAP), together form an innovative mix of mechanisms that have promoted compliance among parties.

A second noteworthy feature of the Montreal Protocol is its decision-making procedures. According to Article 11, the first meeting of parties in Helsinki in May 1989 decided that decisions on matters of substance were to be made by a two-thirds majority vote and decisions on matters of procedure were to be made by a simple majority vote of Parties present. This is in accordance with rule 40 in the Rules of Procedure for Meetings of the Conference of the Parties to the Vienna Convention and Meetings of Parties to the Montreal Protocol (UNEP, 2000). Thus, in comparison to most other international agreements, decisions can be made by a majority of parties; i.e. either a two-thirds majority for substantive issues or a simple majority for procedural issues. This reduces the opportunity for minority vetoing.

The third feature that is particular to the Montreal Protocol is its rapid goal attainment. Only eight years elapsed from the time the first binding agreements were made in 1987, to 1995 when CFCs and halons were almost entirely phased out in most OECD countries. How could an international agreement lead to such effective results in such a short time? The success is mainly related to the problem itself. Reducing the production and consumption of ODS during the 1990s was regarded as far more feasible than it was in the 1980s (Oberthür, 1999). When science was able to provide proof of the damage to the ozone layer from the release of ODS into the atmosphere and industry saw the potential for a large worldwide market in substitute chemicals, interests became consensual rather than conflicting as they had been during the 1980s, and effective results were reached rapidly.

The Norwegian ozone policy and linked international regimes. Norwegian ozone policy has not emerged in a vacuum, and this section addresses the role of other processes and agreements that have affected ozone policy. At the international level there are mainly two regimes that have affected Norwegian ozone policy: the 1979 Convention on Long-Range Transboundary Pollution (CLRTAP), and the 1947 General Agreement on Tariffs and Trade (GATT), replaced by the World Trade Organisation in 1995 (WTO).

During the 1970s and 1980s, Norwegian environmental policy focused to a large extent on problems related to transboundary air pollution, later regulated under the CLRTAP (see Chapter 5). This agreement mainly regulates substances affecting the troposphere, while the Montreal Protocol addresses substances affecting the atmosphere. The CLRTAP was signed in 1979 and came into force in 1983, thus becoming one of the first environmental treaties ratified and valid globally. The Nordic states played an influential role in pushing for an agreement, which had a spillover effect, affecting Nordic positions and participation in ozone negotiations. Norwegian participation in CLRTAP was a learning experience and affected the way Norway pursued national goals in the ozone negotiations. Particularly seen from a scientific perspective, Norway was a leading state, and scientists were allowed to take part in official Norwegian delegations to both the CLRTAP and the Montreal Protocol. This obviously affected Norwegian positions under both treaties (Bakken, 2001; Thompson, 2003).

The conflict between the goals of environmental protection and trade liberalisation has been discussed intensively over the last decade, particularly with respect to the Dolphin-Tuna case between Mexico and the US. There have been no rulings or major incidents with respect to the relationship between the WTO (former GATT) and the Montreal Protocol, but the restriction on trade in both treaties means that there is a potential for confrontation (or synergy). As the Montreal Protocol contains measures either banning or regulating trade, some have feared that states not complying with trade measures may claim the protection of WTO to avoid being sanctioned.[6] But as a matter of fact, the GATT Secretariat already in the early phases of negotiations on the Protocol stated that such bans were consistent with trade rules and there have not been any incidents brought before the WTO tribunal (Benedick, 1998). In other words, potential disruptive WTO reactions to trade bans and trade limitations agreed on certain substances under the Montreal Protocol have not materialised, and thus have not affected Norwegian goal attainment.

In general, there have not been many problems related to the interaction with other linked regimes, and neither Norway's goal attainment nor the goal attainment of the core regime seems to have been altered significantly. Bull, the official responsible for monitoring compliance in Norway in the PCA, comments that:

> The importance of complying with the targets set under the Montreal Protocol and the achievement of other national objectives under the ozone policy has not depended on, or been obstructed by, goal attainment within other issue areas.[7] (citation from interview with Bull in Vaggen Malvik, 1998, p.68)

This is mainly related to the fact that the Montreal Protocol and ODS were among the first environmental problems to be addressed by the international community, and to the fact that ozone depletion is a benign problem compared to other issue areas such as climate change, air pollution, and water pollution.

The National Dimension

The effects of the Montreal Protocol and other international processes on goal attainment in Norway. In what way did the Montreal Protocol affect domestic implementation of ozone policy in Norway? Initially, Norway's influence on the protocol must be deemed more influential than the protocol's effect on Norway. Norway was among the countries that pushed for stricter regulations and a quicker phase-out. Thus, in this first period, until the protocol came into force towards the late 1980s, Norwegian goals and policies affected the international regime. However, later on, once established, the protocol and subsequent amendments throughout the 1990s have developed beyond Norwegian goals, becoming an important framework for the cooling industry in Norway, and thus also for Norwegian ozone policy. In recent years a trend seems to have been reinforced where new policy developments have gradually been imposed on Norway from abroad. Thus, over time, the protocol has become more important for goal attainment also in Norway. The protocol has broad support in the international community; strict, clear, and legally binding objectives; a high degree of specificity; and better compliance and enforcement procedures than most other environmental treaties. Together these factors have made the Protocol an effective international treaty, which also benefits Norwegian environmental policy.

Altogether, there is no reason to believe that the Montreal Protocol as an independent factor contributed significantly in facilitating Norway's goal attainment, although it obviously inspired the formulation of the 1988 Action Plan, which led to an agreement between industry and government in 1991 to phase out the consumption of ODS in Norway. Norway did not produce ODS, was not a large consumer of ODS, and had access to so-called natural cooling substances that did not contain ODS. The costs involved for Norway were low due to the lack of production facilities, and the benefits were high due to Norway's geographical location in the Northern Hemisphere where the effects of ozone depletion would be more severe than around the equator. Thus, the domestic situation implied that opposition to a phase-out was not significant because industry and consumers did not rely as heavily on ODS as they did in other countries. In addition, the government was active in funding research on ODS substitutes, in making regulations transparent, and in diffusing information; an approach which has proven to be effective (Vaggen Malvik, 1998). Phase-out schedules were generally stricter and introduced earlier than was the case internationally. Thus, the combination of beneficial circumstances and political pressure from the government seems to have contributed more to goal attainment in Norway than what can be attributed to the Montreal Protocol. Clearly, without the Montreal Protocol, there is reason to believe that efforts in Norway would have been less effective.

But in general, goal attainment owes more to particular national circumstances than to the core international agreement on the subject (Bakken, 2001).

However, apart from the Montreal Protocol, it is worth mentioning a few other agreements that have contributed to Norwegian goal attainment. During the 1970s and 1980s, sea pollution became a major issue for Norway. Norway's long coastline and the rapid development of offshore oil exploration were reflected in environmental awareness around questions related to sea pollution, and Norway participated in establishing the Convention for the Protection of the Marine Environment of the North East Atlantic of 1992 – better known as the OSPAR regime (see Chapter 6). This agreement resulted in an overlap of substances covered: Carbon tetrachloride and trichloroethane are both solvents with ozone-depleting potential and multiple modes of application, as well as being toxic and carcinogenic, particularly when exposed to aquaculture. Thus both substances are regulated by the Montreal Protocol (as ozone depleters) and the OSPAR (as toxic carcinogenic chemicals exposed in water). As both agreements aim to phase out these two substances, the interaction has been synergetic, supporting Norwegian goal attainment (Vaggen Malvik, 1998).

The EEA agreement requires Norway to abide by any decision or agreement made by the EU concerning the single market.[8] This is also valid for the EU ozone regulations. EU regulations have affected Norway by improving the environmental standards within Europe and thereby reducing the potential threat to the ozone layer and the effect on Norway. During the early 1990s, this implied that a number of EC Member States were forced by a majority to agree on faster phase-out schedules than they otherwise would have unilaterally implemented (Rowlands, 1998). Even though these decisions did not affect Norway's policy directly (apart from covering a few more substances that were of no particular importance in Norway), they affected progress on protecting the ozone layer, thereby facilitating Norwegian overall goal attainment of resolving a *global* problem faster. The EU has recently agreed on a new and comprehensive regulation, 2037/00/EC, which will have implications for Norway because it aims at phasing out HCFCs and methyl bromide faster than what was agreed to under the Montreal Protocol. In general, Norwegian regulations have been stricter as they have contained restrictions on production, import/export *and consumption*, whereas EU regulations have only targeted production and import/export. But from October 2000, the EU has also restrictions on imports and consumption that will be stricter than existing Norwegian legislation.

With regard to the UNFCCC of 1992 and the Kyoto Protocol of 1997, there has been an overlap in terms of gases covered by the treaties that has to some extent been problematic. To help phase out ODS under the Montreal Protocol, parties have been encouraged to adopt substitute substances such as HCFCs and HFCs. HFCs in particular, but also HCFCs as well, are powerful greenhouse gases, much more potent than carbon dioxide (CO_2) and are thus regulated by the Kyoto Protocol (Oberthür, 2001).[9] This suggests a conflict of interest between the two regimes. However, their common interests become more evident if we consider that while greenhouse gases (including HFCs and HCFCs) warm the atmosphere, they actually have a cooling effect on the stratosphere where the ozone layer is formed. This cooling prolongs, enhances and intensifies ozone depletion, causing a delay in its recovery (ibid., p.360). Thus the very substances that were recommended to preserve the ozone layer actually contribute to its depletion in the long run. On the other hand, there are a number of factors implying that future problems related to consumption of HCFCs and HFCs will not be insurmountable. The consumption of HCFCs is supposed to end in 2010 in Europe (EC Regulation No 2037, 2000), while production may continue for essential uses and export to developing countries until 2025. The consumption of HFCs is projected to grow considerably towards 2010, but as HFCs are energy efficient, and as there are viable substitutes for HFCs, most agree that the HFC problem will be resolved (IPCC, 2000). From the Norwegian point of view, global growth in the consumption of powerful greenhouse gases such as HCFCs and HFCs will delay the recovery of the ozone layer, but seem to cause more problems for climate change than the other way around as suggested by Oberthür, who states that 'while HFCs have been considered part of the solution under the Montreal Protocol, they are part of the problem under the Kyoto Protocol' (Oberthür, 2001: 362).

Domestically, the consumption of HCFs is estimated to grow considerably and is estimated to reach approximately 3 per cent of the total emissions of GHGs in Norway by 2010 (PCA, 2001a).[10]

With respect to HFCs, the Pollution Control Authority has started to look for other substitute chemicals that may replace them, as well as considering how to reduce their consumption. In a recent report, the Pollution Control Authority advised the Ministry of the Environment to implement a tax on the import and consumption of HFCs, a proposal that was included in the 2003 national budget. The tax on the imports of HFCs came into force in January 2003. Thus, due to the tax and the availability of natural cooling agents such as ammonia, propane, butane and hydrocarbons, which neither harm the ozone layer nor have any greenhouse effect, the Norwegian government will most probably manage to reduce the potential growth in the use of HFCs (Ministry of the Environment, 1995, p.4; Ministry of the Environment, 1997-98).

The national political process: *fortunate circumstances and effective efforts*. The Norwegian government had various legal, economic and administrative means at its disposal to phase out existing import and consumption of ODS in Norway. The first regulation banning the use of CFCs in spray-cans was in effect as early as 1981, which led to a 50 per cent reduction of CFC use in spray-cans before 1983. Seven years later, in 1988, the government formulated the first Norwegian Action Plan signaling the coming into force of a tax on CFC use. One year later the Parliament decided that the tax was to be implemented from July 1990.

The affected sectors and businesses – dry-cleaning, cooling, plastics and electronics producers – reacted by forming an association of CFC users called KBF (*KFK-Brukernes Fellesutvalg*). After pressure from KBF, the tax was made conditional on the target groups *not* reaching required reduction targets within assigned limits, and the coming into the force of the regulation on ODS was delayed from July 1990 until 1991. Thus, the proposed regulation of ODS was implemented more or less as a voluntary agreement between government and industry, with the threat from the government that a tax would be implemented if industry did not reach set targets. As the consumption of CFCs dropped by 66 per cent in the period 1986-1991, compared to the required reduction of 50 per cent, the tax did not come into force. Thus the voluntary agreement between the government and industry, combined with the threat of implementing a tax, was effective (Vike, 2001). In the years following 1991, under the threat of introducing a tax, KBF saw no other solution than to cooperate with environmental authorities to enforce the agreement adopted in 1991. KBF received funds to inform its members and made active use of the agreement to persuade its members to comply with the official phase-out schedule (Vaggen Malvik, 1998).

Another reason for KBF to cooperate with the government in phasing out CFCs and halons was the possibility of giving Norwegian industry a competitive edge vis-à-vis foreign competitors in the development of ODS-free refrigeration and dry-cleaning solutions. Norway's traditional dependence on natural cooling substances provided the chemical industry with a competitive advantage in producing ODS-free solutions for the market. With economic support from the government, the industry actually succeeded in innovating new solvents and fire fighting equipment that could replace the old, ozone-depleting substances (Ministry of the Environment, 1995). Norway was among the first countries to impose restrictions, and when the agreements made under the Montreal Protocol turned out to be similar to the Norwegian ones, this gave Norwegian industry opportunities to gain market shares (Bakken, 2001). Later, KBF's role diminished as the import and consumption of CFCs dropped effectively in line with the 1991 agreement, and the organisation was dissolved in 1996.

CFCs and halons were phased out from 1 January 1995, and carbon tetrachloride and trichlorethan were phased out from 1 January 1996. Of the six original substances covered by regulations, only two remained: methyl bromide and HCFCs. The use of methyl bromide was reduced by 56 per cent from 1991 to 2000, and has been reduced further from 4.4 tons consumed in 2000 to four tons in 2001 (PCA, 2001a). This reduction took place to fulfil new EU targets according to 2037/00/EC of a 60 per cent phase-out by 2001. Because there are substitutes available and the use of methyl bromide is of minor importance in Norway, phasing out the remaining four tons will most certainly not be a problem (Søyland, 2001).

HCFCs are also regulated by Regulation 2037/00/EC, and are supposed to be phased out gradually towards 2010.[11] In 2000, 936 tons of HCFCs were imported for consumption in Norway, which represented a two per cent reduction compared to 1999 levels. HCFCs are mainly used as a cooling agent for cold storage purposes in refrigerators and freezers, for producing certain types of insulating foam, and for air conditioning in automobiles. The Pollution Control Authority has estimated that the costs involved in phasing out HCFCs by 2010 are quite large, and account for a major share of the NOK 290-470 million that have been estimated as implementation costs for adopting Regulation 2037/00/EC in Norway.[12] The insulation industry consumed approximately 60 per cent of HCFCs imported to Norway in 2000. But as the industry has planned to phase out HCFCs in 2001, implementation of HCFC regulations will not cause problems for this sector. The remaining consumption of HCFCs is mainly used by the Norwegian fishing fleet for refrigeration purposes. As approximately 95 per cent of the fishing fleet uses HCFCs for refrigeration, this industry will have to bear the costs incurred by a phase-out, estimated to be about NOK 60-100 million (Haukås, 2000).

Besides HCFCs, there are also a few 'new substances' entering Norway that the Pollution Control Authority has considered banning in terms of both import and consumption (PCA, 2001a). So far there are two substances that will be targeted: N-propyl bromide and chlorobromomethane. Their ozone-depleting value and effect are still under study, but the EU proposed that they be subject to Montreal Protocol regulations at the Meeting of the Parties in 2001. The EU proposal was watered down by the US and Japan, and it is uncertain whether these substances will be covered by the Montreal Protocol at all (Campbell, 2001). Thus, if a decision is made to implement a ban on new substances in the near future, Norway will be among the first to regulate these substances.

The efforts of the Norwegian environmental authorities to phase out the various types of ODS have met little resistance. Inter-departmental conflicts have not appeared due to the benign character of this environmental problem. Compared with other environmental problems, ozone depletion was for Norway a relatively simple issue to handle, which implied less conflict in connection with policy- and decision-making, and higher effectiveness in connection with efforts to phase-out ODS.

To sum up, Norwegian phase-out efforts have been highly effective judging by the accomplished results. The most important factors behind the phase-out were definitively introducing the regulation and the threat of imposing a tax on ODS. Additional factors contributing to the effectiveness of the phase-out were the active involvement of target groups through information dissemination, cooperation and voluntary agreements, which led eventually to a 98 per cent reduction in domestic consumption of ODS (Vike, 2001). The phase out of the most potent substances, CFCs and halons, was facilitated by early action and signals from environmental authorities to the industry of what was to come, but also by the fact that target groups were few and well organised (refrigerator and dry-cleaning sectors), which facilitated cooperation and information-sharing at an early stage (Vaggen Malvik, 1998). Compared with a global reduction of 85 per cent, and an OECD reduction of 90 per cent, the phase-out of ODS in Norway is clearly a case of over-compliance. On the other hand, it is obvious that Norway had an extremely favourable starting point. The costs involved in switching from ODS to ODS-free substances were low, the target groups were few and well organised, and substitute chemicals such as ammonia, CO_2, propane, and hydrocarbons were easily available.

The Foreign/National Dimension

To what extent have those actors responsible for implementation been included at an early stage? Starting with the internal foreign dimension, we have already stated that affected target groups were included in the decision-making process as early as the late 1980s and onwards. In other words, there has been a high level of coordination between government and industry, which has secured an effective phase-out of ODS consumption in Norway. Providing information to and including consumer groups at an early stage enhanced the target groups' support for policy, leading to an effective phase-out (Vaggen Malvik, 1998). The positive stance taken by industry was obviously related to the availability of government support and substitute chemicals that allowed refrigerator and dry-cleaning sectors to continue their operations (Søyland, 2001).

On the governance side, the same environmental authorities have been in charge of implementation and policy development from the beginning. A lot of the operative responsibilities and the day-to-day work have been left with the Pollution Control Authority, which participates in EU expert committee meetings. The agency formulates new regulations, gathers opinions and is responsible for contacting and hearing target groups in association with the development of new regulations (Bull, 2001). The Ministry of the Environment participates in high-level meetings held yearly under the Montreal Protocol, and follows closely the political development at the EU level. Thus, its role is mainly to assess the degree to which Norwegian policy needs to be updated in response to developments at the international level. Altogether the institutional set-up does not create problems for further problem-solving within the ozone policy area. Responsibility is well divided between the Ministry and the Pollution Control Authority, and this distribution does not create any problems for further problem-solving and policy-making with respect to the smaller problems that remain within the area.

Coordination between efforts at the international and national levels has been particularly effective in Norway's case, and facilitated a high level of goal attainment. However, even though the problem is more or less solved in Norway, there are still a few problems remaining and several developing countries are still struggling to fulfil their obligations in accordance with the Montreal Protocol.

Alternative Explanatory Factors

Norway has enjoyed a favourable situation from which to pursue ozone policy internationally and domestically. First, at the domestic level, Norway's dependence on ODS has been relatively minor compared to other states because natural cooling agents were already being used in larger refrigeration plants, implying that the cost involved in the change from ODS to ODS-substitutes has been low (Søyland, 2001). In 1986, total consumption of CFCs was 1400 tons a year in Norway, compared to 5300 tons in Sweden, 800 tons in Denmark, and 3000 tons in Finland. Thus, Norway has not been particularly dependent on CFCs, and with considerable scientific and technical expertise in the development and use of ODS-free substances, alternative substances have been marketed for use in the refrigeration, air-conditioning and heat pump equipment sectors (ibid.).

Second, the lack of any major opposition, such as industry associations, or trade interests, to regulating CFCs made it easier for Norway to assume a leadership role (Søyland, 2001; Bull, 2001). The cooling industry and other sectors that were dependent on ODS saw regulations as a means to improve their position in the growing market of ODS-free substitutes, rather than as something that would distort their market positions. The costs involved for behavioural change within targets groups was obviously an important factor here. Environmental authorities estimated that phasing out CFCs, halons and other ODS would cost society approximately NOK 65 million a year (Ministry of the Environment, 1995). In comparison, the same study estimated the costs involved in cleaning up the Norwegian industry's use of toxic substances and other material to amount to approximately NOK 200-400 million (ibid.). The costs of cleaning up the environment by reducing pollution at sea, emissions to air and the environmental consequences of agricultural activities can be estimated to be considerably higher, but are more difficult to estimate, as the number of actors and sectors involved are numerous. Thus, phasing out ODS involved comparatively lower costs for industry than what has been associated with environmental problem-solving in most other cases. In other words, the government and the industry had common interests in strengthening the regulatory regime domestically and internationally, which facilitated Norway's proactive policies on reducing the consumption, production and trade of ODS at both levels.

Internationally, Norway was part of a coalition consisting of powerful states such as the US and Canada, which were supported by UNEP and its Executive Director, Dr Mostafa Tolba. Without support from the world's largest producer of knowledge about the atmosphere and of substances harming it, the US, the coalition would probably not have achieved much during negotiations. In other words, it is only as a member of the Toronto group that Norway has achieved positive results internationally.

Together, these factors facilitated an effective Norwegian ozone policy that more or less solved the problem as perceived from a Norwegian perspective (Søyland, 2001). The approach chosen by the government where clear guidelines for industry and target groups were set at an early stage, combined with financial incentives and voluntary agreements, has been particularly successful and a model for future environmental problem-solving (Vaggen Malvik, 1998, p.103). This statement has also been supported by major environmental NGOs in Norway such as the Norwegian Society for the Conservation of Nature (*Norges Naturvernforbundet*) (ibid.). The benign character of the problem has made ozone depletion a special issue that is not easily compared with larger issues such as climate change, air pollution and water pollution. Furthermore, the international drive, both within the EU and under the Montreal Protocol, to further develop ozone policy to cover new substances and set tougher targets has led to a constantly evolving policy at the national level. In conclusion, effective national efforts undertaken by environmental authorities and industry in tandem coupled with an international drive for higher goal attainment has made Norwegian phase-out of ODS a straightforward success story.

Conclusion

How could Norway reach its environmental goals so easily within this area? The answer lies mainly with the beneficial circumstances for problem-solving in this case at both the national and international levels. The positive interplay between policy instruments, good governance, and the benign character of the problem at hand are mostly to credit for the success. Norwegian negotiators and officials have been able to influence the contents of international as well as national regulations of ODS, essentially because they faced little resistance. At the international level, the US u-turn in 1983 and the discovery of the hole in the ozone layer over the Antarctic in the mid 1980s were crucial in changing the political atmosphere with respect to regulating ODS production and consumption. At the national level in Norway, the minimal consumption of ODS, the lack of any major industry opposition to regulations, and the selection of policy instruments by environmental authorities are probably among the major explanatory factors behind the rapid phase-out. Norwegian industry benefited from an early phase-out, which facilitated the progressive stance taken by the Norwegian government. However, the achievements at the national level were facilitated by the Montreal Protocol; without the protocol it is doubtful whether Norwegian industry would have accepted phasing out their use, because it might have put them at a competitive disadvantage compared to other countries not subject to similar regulation. Thus, the Montreal Protocol facilitated domestic action, but it was the cooperation between Norwegian government and industry that secured the effective phase-out of ODS in Norway. Through cost-effective means, Norway reduced its consumption of ODS by 98 per cent, and has managed to solve its share of the 'ozone problem'.

Other institutions such as OSPAR, CLRTAP and the EU have created a positive crossfire, promoting goal achievement in Norway. Thus, the only coordination problem of any significance that is still valid as a problem today is the use of the substitute gases HFCs and HCFCs, which is causing problems for Western countries' efforts to reduce greenhouse gas emissions. In conclusion, Norwegian goals within the field of ozone layer protection have been effectively served both internationally and domestically: internationally through the Montreal Protocol, and domestically through the cooperation between the Pollution Control Authority and the Ministry of the Environment. At the time of this writing, the ozone layer is seen as relatively unproblematic in Norway, and the EU and the Norwegian environmental authorities seem to be handling remaining challenges well. A recent drop of 33 per cent in the total import of ODS from 2001 to 2002 and the implementation of a Norwegian tax on the import of HFCs from 1 January 2003 proves that the environmental authorities are in good control of remaining challenges (Ministry of the Environment, 2003). The handling of these issues has attracted little or no political attention, which is yet further proof of the lack of controversy associated with the reduction in imports and consumption of ODS to Norway.

Notes

1 See Parson et al. (1995) for a comprehensive discussion of the many ODS covered by international ozone agreements.

[2] In 1983 Canada, Finland, Norway, Sweden, and Switzerland formed a coalition named after the city where these countries held their initial meeting. The Toronto group introduced the idea of reducing CFC emissions to the international community. The US joined the coalition in late 1983 (Benedick, 1991, p.42).

[3] The EEA agreement entered into force in 1994. After 1994, regulations, decisions and policies related to the EU single market, which also include the EU ozone regulations, apply in Norway directly. This is according to Article 6, and specifically according to Annex XX, ch. III. Air, in the Agreement on the European Economic Area.

[4] Phasing out the remaining use of HCFCs by 2010 will cost about NOK 290-470 million, according to the Norwegian Pollution Control Authority (PCA, 2001b).

[5] See Annex V of the report of the Fourth Meeting of the Parties, Montreal Protocol.

[6] The Montreal Protocol's Article IV contains measures aimed at: a) banning trade between Parties and non-Parties of the Protocol to reduce the incentive to increase production in or move production to non-Parties, b) banning trade in certain substances between Parties to avoid illegal trade and c) securing a system of licensing imports and trade in ODS also to reduce the potential for illegal trade.

[7] Author's translation from Norwegian.

[8] In general according to Article 6, and specifically regarding environmental issues according to Annex XX, ch. III. Air, in the Agreement on the European Economic Area.

[9] HFCs are approximately 1200 to 3200 times more potent than CO_2 in terms of their contribution to the greenhouse effect (Vaggen Malvik, 1998, p.75).

[10] In 2001 HFCs and perfluorocarbons (PFCs) together represented approximately 0.45 per cent of Norway's total GHG emissions in 2001 (Ministry of Finance, 2002).

[11] The schedule according to Regulation 2037/00/EEC is stabilisation from 1 January 2001 at a level equaling consumption at 1989 levels plus 2 per cent of CFC consumption in 1989 measured in ODP values. Towards 2010, Norway is expected to reduce its consumption by 15 per cent from 1 January 2002, 55 per cent from 1 January 2003, 70 per cent from 1 January 2004, 75 per cent from 1 January 2008, and 100 per cent from 1 January 2010 (PCA, 2001a).

[12] These costs cover the expenses of phasing out all regulated ODS in a ten-year period until 2010.

References

Bakken, P. (2001), Personal Communication, 30. November 2001. Former Head of Norwegian Delegation to the Vienna Convention and the Montreal Protocol. Current position is: Deputy Director, Division of Technology, Industry and Economics, UNEP, Paris, France.

Benedick, R. (1991), *Ozone Diplomacy: New Directions in Safeguarding the Planet*, Harvard University Press, London.

Benedick, R. (1998), *Ozone Diplomacy: New Directions in Safeguarding the Planet*, 2nd Edition, Harvard University Press, London.

Bull, A.-M. (2001), Personal Communication, 16. November 2001. Senior Advisor, Section on Chemicals and Chemical Industry (NKK), State Pollution Control Authority (PCA) (Statens forurensningstilsyn, SFT), Oslo.

Campbell, N. (2001), Personal Communication, 13. November 2001. Environment Manager, ATOFINA SA (formerly Elf Atochem).

CGCP, Canadian Global Change Program (2002), 'Lessons from the Montreal Protocol', The Canadian Global Change Program (CGCP) seminar in association with the 10th Anniversary of the Montreal Protocol, September 13, 1997, retreived 13 March 2002 from www.globalcentres.org/cgcp/english/html_documents/ads/coll-e.htm.

Dahl, A. (1999), 'Miljøpolitikk – full tilpasning uten debatt', in D.H. Claes and B.S. Tranøy (eds), *Utenfor, annerledes og suveren? Norge under EØS-avtalen*, Fagbokforlaget, Bergen, pp. 127-47.

EC Regulation No 2037 (2000), of the European Parliament and the Council of 29 June 2000 on Substances that Deplete the Ozone Layer, Official Journal L 244/1, 29/9/2000, pp. 0001-0024.

Gehring, T. (1994), *Dynamic International Regimes: Institutions for International Environmental Governance*, Peter Lang GmbH, Frankfurt am Main.

Haukås, H.T. (2000), 'Cost Estimates Related to Strenghtened HCFC –Regulations', study commissioned by the Norwegian State Pollution Control Authority (PCA) (Statens forurensningstilsyn, SFT), Oslo.

IPCC, International Panel on Climate Change (2000), *Emissions Scenarios: A Special Report of IPCC Working Group III*, published by UNEP and the WMO, retreived 13 November 2001 from www.grida.no/climate/ipcc/emission/index.htm.

Ministry of the Environment (1989), *Environment and Development. Programme for Norway's Follow-up of the Report of the World Commission on Environment and Development* (Miljø og Utvikling. Norges oppfølging av Verdenskommisjonens rapport), Report No. 46 to the Storting (1988-89), Ministry of the Environment, Oslo.

Ministry of the Environment (1995), *Types of Measures in Environmental Policy-making* (Virkemidler i miljøpolitikken), NOU report Nr. 1995:4, Ministry of the Environment, Oslo.

Ministry of the Environment (1997-98), *Norway's Follow-up of the Kyoto Protocol* (Norges oppfølging av Kyotoprotokollen), Report No. 29 to the Storting, Ministry of the Environment, Oslo.

Ministry of the Environment (2003), 'Miljøvernminister Børge Brende glad for reduksjonen i forbruket av ozonreduserende stoffer', Press release retreived 14 April 2003 from http://odin.dep.no/md/norsk/tema/ forurensning/arkiv/022051-070084/index-dok000-b-n-a.html.

Ministry of Finance (2002), 'Spørsmål nr. 118, fra Finanskomiteen/ Arbeiderpartiets fraksjon, av 28. oktober 2002, vedrørende Statsbudsjettet 2003' Ministry of Finance, 02/3921, retreived 6 December 2002 from http://odin.dep.no/fin/norsk/ Korrespondanse/bud2003/part/ap/006041-990883/index-dok000-b-n-a.html.

Molina, M.J. and Rowlands, F.S. (1974), 'Stratospheric Sink for Chlorofluormethanes: Chlorine Atom-catalysed Destruction of Ozone', *Nature*, vol. 249, pp. 810-12.

Oberthür, S. (1999), 'The EU as an International Actor: The Protection of the Ozone Layer', *Journal of Common Market Studies*, vol. 37, no. 4, pp. 641-59.

Oberthür, S. (2001), *Production and Consumption of Ozone Depleting Substances 1986-1999: The Data Reporting System under the Montreal Protocol*, GTZ, Escborn.

Parson, E.A. and Greene, O. (1995), 'The Complex Chemistry of the International Ozone Agreements', *Environment*, vol. 37, no. 2, pp. 16-43.

Porter, G. and Brown, J.W. (2000), *Global Environmental Politics*, 3rd edition, Westview Press, Boulder, CO.

PCA, Norwegian Pollution Control Authority (Statens forurensningstilsyn, SFT) (2001a), 'Miljøstatus i Norge: Ozonlaget', retreived 18 October 2001 from www.miljostatus.no/Tema/Klimaluftstoy/Ozon/ozonlaget.stm.

PCA, Norwegian Pollution Control Authority (Statens forurensningstilsyn, SFT) (2001b), Forskrift om ozonreduserende stoffer – Høring, a letter from PCA to consumers concerning the implementation of Regulation 2037/00/EC in Norway, 28 January 2001.

Rowlands, I. (1998), 'EU Policy for Ozone Layer Protection', in J. Golub (ed.), *Global Competition and EU Environmental Policy*, Routledge, New York.

Søyland, S. (2001), Personal Communication, 29 June 2001. Senior Executive Officer, Ministry of Environment, Oslo, Norway.

Thompson, J. (2003), Personal Communication, 5. November 2003. Former Principal Officer in the Ministry of the Environment with responsibilty for the Montreal Protocol and the CLRTAP in the 1980s. Current position is: Secretary General of the Norwegian Mountain Touring Association (DNT).

UNEP (2000), *Handbook for the International Treaties for the Protection of the Ozone Layer*, 5th edition, UNEP Ozone Secretariat, Nairobi.

Vaggen Malvik, H. (1998), 'Implementering av Montrealprotokollen i Norge – Grad av oppfølging og forklaring på grad av oppfølging av Montrealprotokollen, sett i lys av en R-modell og en DP-modell', Hovedoppgave statsvitenskap, University of Oslo, Institute for Political Science, Oslo, Autumn.

Vike, E. (2001), Personal Communication, 30. November 2001. Head of Section on Chemicals (NKV), State Pollution Control Authority (PCA) (Statens forurensningstilsyn, SFT), Oslo.

WMO (1999), *Scientific Assessment of Ozone Depletion: 1998*, World Meteorological Organisation Global Ozone Research and Monitoring Project – Report 44, World Meteorological Organisation, Geneva.

Chapter 5

Air Pollution: International Success, Domestic Problems[1]

Jørgen Wettestad

Introduction

A natural starting point for the history of the international politics of transboundary air pollution is the Swedish scientist Svante Oden's paper 'The acidification of air and precipitation and its consequences in the natural environment', published in 1968. Oden argued that precipitation over Scandinavia was becoming increasingly acidic, damaging fish and lakes. Moreover, he maintained that the acidic precipitation was largely caused by sulphur compounds from British and Central European industrial emissions. This paper did not go unnoticed in Norway and can be seen as a starting point for a Scandinavian political campaign to place transboundary air pollution higher on the international agenda and establish an international institution to deal specifically with these problems.

As early as in 1977-78 it was becoming well substantiated – at least with regard to sulphur dioxide (SO_2) emissions – that European emissions were of far greater importance than domestic emissions for environmental conditions in Norway. Hence, Norway and Sweden acted as central pushers in the negotiations leading up to the 1979 Convention on Long-Range Transboundary Air Pollution (CLRTAP) within the context of the United Nations Economic Commission for Europe (UNECE). Although the pushers faced reluctance and opposition from important actors such as the United Kingdom and the Federal Republic of Germany, a framework convention was established with the overall goal to 'endeavour to limit and, as far as possible, gradually reduce and prevent air pollution, including long-range transboundary air pollution' (Article 2). (Chossudovsky, 1989; Gehring, 1994). And despite the lack of legally binding emission reduction obligations, in this case there was – at least initially – an almost perfect match between regime objectives and national Norwegian goals.

The international chapter of the Norwegian story is then a highly successful one. International commitments covering the most important pollutants have been established, and these commitments have over time become both stronger and 'smarter', i.e. more differentiated and sophisticated. European emissions have been significantly reduced. However, international success has not been matched by equally successful domestic policies. True, the domestic reduction of SO_2 emissions is a shining success story. But Norway's performance with regard to reducing emissions of nitrogen oxides (NO_x) and volatile organic compounds (VOC) is much less impressive. In fact, the VOC performance is simply a fiasco. The same holds true for air quality in general, with the exception of the reduction in SO_2.

In order to account for Norway's varying and partly embarrassing goal attainment, this chapter emphasizes a combination of domestic institutions and policies and fundamental problem characteristics. One common factor in particular that has complicated the development of effective NO_x and VOC policies is clearly the growth of Norway as an oil and gas producer and exporter in the 1990s. In order to deal effectively with this challenge, the increasing linkages and complementarity between CLRTAP and EU policies may provide the environmental authorities with the necessary clout to get adequate policy packages adopted and implemented in the years ahead.

Norway's Goal Attainment in the Field of Air Pollution After 1979

Background: An Overview of Goals and Commitments

As indicated, the overriding goal for Norway in this issue area from the very beginning has been to reduce transboundary and domestic emissions contributing to acidification and related environmental damages – with a primary emphasis on the transboundary emissions. For instance, government White Paper no. 44 (Ministry of the Environment, 1975) states that the government will seek to reduce European SO_2 emissions (ibid., p.117). Focus was specifically on SO_2, and prime emphasis was given to improving knowledge on long-range transport of pollutants and the effects of this type of pollution compared to air pollution of local origin.

At subsequent stages, this overriding goal was specified and elaborated. The 1984 White Paper on water and air pollution policy measures (Ministry of the Environment, 1984) expresses the goal of reducing SO_2 emissions by at least 30 per cent by 1993 in relation to 1980 levels. It also puts forward the intention to establish a 50 per cent reduction by 1993 as an international target (ibid., p.8). The 1985 CLRTAP Helsinki Protocol then adopted the common goal of reducing SO_2 emissions by 30 per cent, relative to a 1980 baseline, by 1993.[2] However, as indicated, Norway's ambitions were higher than that, and the 1989 White Paper on environment and development established a 50 per cent reduction by 1993 as the national goal for Norway (Ministry of the Environment, 1989, p.60).

However, transboundary air pollution damages do not stem from SO_2 alone. A Norwegian draft related to the work on the Helsinki Sulphur Protocol expressed a need to reduce European NO_x emissions and recommended including quantified, substantial reductions of emissions or of transboundary fluxes of nitrogen in the protocol. It is reasonable to assume that this also was the official Norwegian position when the NO_x pre-negotiations started in October 1985 (Stenstadvold, 1991, p.87). However, when the main negotiations started in 1988, the Norwegian goal had turned into a more modest call for stabilisation of emissions by 1994 (Laugen, 1995; Wettestad, 1998). Along with the other CLRTAP parties, this commitment is also what Norway agreed to in the 1988 Sofia Protocol.[3] Moreover, it is interesting to note that Norway on the same occasion also reluctantly signed a political Declaration calling for 30 per cent emission cuts by 1998 (see below). In the domestic process that followed, this 30 per cent target was established as a national goal.[4]

Furthermore, the 1988-89 White Paper on environment and development and Norway's follow up to the World Commission for Environment and Development (WCED) specified two international goals: to use critical loads in the environment as the basis for subsequent international air pollution agreements, and to establish international agreements on reducing emissions of VOCs (Ministry of the Environment, 1989, p.46).[5] A couple of years later, the goal of establishing an agreement on VOCs was further specified in connection with the negotiations on a VOC protocol under CLRTAP. Norway's position in the VOC negotiations emphasized the principles of basing agreements on critical loads and cost effectiveness. As a first step in this direction, Norway (along with Canada) sought and got acceptance for the Tropospheric Ozone Management Area (TOMA) concept, where areas within countries not part of the TOMAs were exempted from commitments.[6] The 1991 CLRTAP Geneva Protocol resulted in Norway agreeing to, first, reduce VOC emissions within the Norwegian TOMA south of the 62nd parallel by 30 per cent by 1999. Moreover, Norway agreed to freeze VOC emissions for the country as a whole.[7] Hence, the VOC Protocol introduced a very rudimentary first differentiation in CLRTAP policy-making.

In line with the understanding that the 1985 Sulphur Protocol was only a first, crude step, a second round of negotiations on reductions of SO_2 emissions took place in the first part of the 1990s. As in the previous SO_2 process, Norway adopted something of a vanguard position. The resulting 1994 CLRTAP Oslo Protocol can be seen as a milestone in the process of differentiating CLRTAP commitments between the different parties on the basis of the critical loads concept.[8] The protocol gave Norway a target of reducing SO_2 emissions by 76 per cent, in relation to a 1980 baseline, by 2000.[9]

From the mid 1990s, in line with the growing understanding of how various pollutants interact, and the related need to adopt a more comprehensive approach to policy-making, preparatory work and negotiations on a multi-pollutant and multi-effects protocol were conducted (Wettestad, 2002). The positions Norway took in these negotiations reiterate the earlier policy profile of Norway: eager to reduce domestic emissions of SO_2, more reluctant to reduce other substances such as NO_x and VOCs. In essence, the 1999 Gothenburg Protocol establishes an unprecedented complex web of differentiated national emission ceilings.[10] This protocol requires Norway to reduce emissions of SO_2 by 58 per cent, NO_x by 28 per cent, and VOCs by 37 per cent, as well as to stabilize ammonia (NH_4). All the reductions are relative to a 1990 baseline and must be achieved by 2010.

Except for the goal of basing international agreements on critical loads, all the goals and commitments summed up so far deal with the reduction of *emissions*. Up until the negotiations on the second CLRTAP Sulphur Protocol, which gave greater weight to effects and critical loads, this was also the predominant policy focus within CLRTAP. In addition to emissions of specific pollutants, however, the policy sub-issue of *air quality* is also pertinent. Air quality has mostly been framed as a domestic and local issue, although a European perspective is also highly relevant in this context. The national goal formulation process in this area started in the beginning of the 1990s. In terms of tracing Norwegian goals, the situation is complicated by the fact that there are three related and overlapping, but not identical, parameters in this context: air quality *criteria* recommended by the Norwegian Pollution Control Authority (PCA) (*Statens forurensningstilsyn* – SFT) and national health authorities; national air quality *goals*, based on both health and socio-economic criteria; and legally binding air quality *limit values*, setting minimum standards (PCA, 2002a). The Pollution Control Authority issued the first set of recommended air quality criteria in 1992. The substances targeted were particulate matter (PM_{10}), NO_2, and SO_2. The standards adopted within the EU context were a central point of reference from the very start. And overall, the criteria recommended by the Norwegian authorities in 1992 were more ambitious than the EU standards.[11]

In 1997, these criteria were tightened (albeit modestly) and turned into legally binding limit values.[12] In addition to the three substances targeted earlier, benzene was added. Except for benzene, for which limit values were immediately operative, the main deadline for compliance with the limits was set for 2005.[13] In addition to the binding 1997 limit values, slightly more ambitious 'strategic national goals' for air quality were established in 1998, targeting the same four substances, mainly with 2010 as the deadline.[14] A new round of revising these standards was initiated in fall 2001, and updated legally binding limit values were then adopted by the Ministry of the Environment in October 2002.[15]

On this background, the following summary picture may be presented:

Table 5.1 Central air pollution goals and commitments

	SO_2 emissions	NO_x emissions	VOC emissions	NH_4 emissions	Environmental conditions
Initial goals	• 50% reduction by 1993 • Intermediate goal: 76% reduction by 2000	• Stabilization by 1994 • 30% reduction by 1998	• 30% reduction within TOMA by 1999 • Freeze of overall emissions by 1999	• No initial goal	• Reduce acidification • Agreements based on critical loads • 1992 air quality recommendations
Revised goals	• 58% reduction by 2010 (Gothenburg Protocol)	• 28% reduction by 2010 (Gothenburg Protocol)	• 37% reduction by 2010 (Gothenburg Protocol)	• Stabilization of emissions by 2010 (Gothenburg Protocol)	• 1997 air quality requirements; in addition, 1998 national goals • 1997 requirements revised and up-dated in 2002

International Goal Attainment: Success on Overriding, International Goals

With regard to the overriding goal of reducing European (transboundary) emissions contributing to acidification and other damages in Norway, goal attainment must be characterised as high. In the period 1980-97, European SO_2 emissions were reduced by around 60 per cent, and sulphur deposition over Norway has been cut in half (Ministry of the Environment, 2001b).

With regard to the international goals which were specified around 1988-89, and the goal of establishing agreements based on critical loads, goal attainment can be characterised as quite high. The 1994 Oslo Protocol took a first, significant step in this direction, although it covered only SO_2. The 1999 Gothenburg Protocol then took a major step further in this direction, by covering several substances and optimizing emission reductions simultaneously between countries and across substances. With regard to acidification, critical loads modelling indicate that critical levels were being exceeded in around 32.5 million hectares of ecosystem area in 1990.[16] In comparison, faithful implementation of the Gothenburg Protocol will reduce this area to around 4.4 million hectares in 2010. However, although this will mean a significant improvement, a substantial gap to critical levels will still remain. With regard to the goal of establishing an international VOC agreement, this was fulfilled by the adoption of the CLRTAP Geneva Protocol in 1991, and then revised VOC commitments in the 1999 Gothenburg Protocol.

Domestic Goal Attainment: Mixed Performance

Turning first to the goals for Norwegian performance established by the initial, single-substance CLRTAP protocols, SO_2 performance has been impeccable. Sulphur dioxide emissions have been reduced by around 75 per cent in the period 1980-97. With regard to NO_x, the picture is more mixed. On one hand, Norway did indeed *initially* comply with the stabilisation commitment in the 1988 Sofia Protocol. In the period 1987-94, NO_x emissions were reduced by six per cent. On the other hand, in the period up to 1998-99, not only did emissions fail to keep dropping, they actually increased – exceeding 1987 levels by 2000 tons in 1999 (Ministry of the Environment, 2000, p.97). Hence, Norway was far from living up to the political declaration goal of a 30 per cent reduction by 1998. And Norway's performance when it comes to VOCs is demonstrably grim. As can be recalled, Norway pledged to both freeze overall emissions and reduce its emissions within the TOMA by 30 per cent by 1999. Instead, not only did emissions not decrease, they actually *increased* by around 20 per cent from 1989 levels.[17] So although after peaking in 1996 emissions have decreased somewhat, this is a clear, outright failure.

With regard to the air quality goals established in 1997 and 1998, the present 'mid-term' status is mixed – although the overall quality is less than satisfactory. Again, the situation is best with regard to SO_2. In the major cities, no one is exposed to SO_2 concentrations above national targets.[18] This is not the case with regard to PM_{10}, NO_2 and benzene. For instance, in Oslo, in the period 1994-98, PM_{10} target levels were exceeded between 35 and 45 times. The national target was exceeded around 20 times in 1998-99 – which was a comparatively good year. In the year 2000, it was estimated that over 80,000 people were periodically exposed to PM_{10} levels above the national target (PCA 2002a). In 2001, 74 days were measured as exceeding the most recent EU NO_2 and PM_{10} requirements. The worst peak was reached in December 2001 (*Dagsavisen*, 2002). So it makes sense that the Pollution Control Authority in February 2003 warned that there is still a considerable way to go to attain national air quality targets (ibid.). But things are not totally bleak. Figures for the period 1995-2001 indicate a 30 per cent reduction in the number of persons in Oslo exposed to PM_{10} levels above national targets. The corresponding trend for NO_2 was a 77 per cent reduction (PCA, 2003).

Summing up, then, Norway has had much success with regard to the international and transboundary aspects of controlling air pollution. Transboundary emissions to Norway, especially SO_2, have been reduced substantially. This is particularly important for Norway. Another important goal for Norway was establishing the use of critical loads in environmental agreements. This was fully realized in the 1999 Gothenburg Protocol, which is the jewel in the crown for the CLRTAP regime – a highly differentiated agreement based on the notion of critical loads. Domestically, goal attainment performance is much more mixed. With regard to the CLRTAP commitments, the SO_2 story is a shining success; NO_x must be characterised as, at best, a medium score; and the VOC performance is a clear and outright failure. With regard to the air quality goals to be fulfilled in 2005, the mid-term situation roughly mirrors emissions reductions: satisfactory for SO_2, unsatisfactory for PM_{10}, NO_2 and benzene. The overall domestic score can then be characterised as mixed, but closer to failure than success.

How can Norway's Goal Attainment be Explained?

International Success: The Result of a Receptive Core Regime and Domestic Coherence?

As indicated in the introduction, there was initially an almost perfect match between national policy ambitions and international regime objectives. In 1979, there was so little doubt that Norwegian policy-makers saw the CLRTAP Convention as the single most important policy instrument to put pressure on major European emitters such as the UK and Germany to bring their emissions down – ultimately reducing acidification and related environmental damages in Norway. Forcefully pursuing the goal of reducing emissions leading to air pollution did not conflict with other Norwegian international environmental policy ambitions – partly because Norway was not participating in many other international environmental regimes at this stage. Of the few regimes Norway was involved in at that time, the only other one that could possibly compete in terms of political attention was the effort to reduce marine pollution, but air pollution was probably perceived as a bigger problem for Norway (see Chapter 6). With regard to linkages between air pollution and other environmental problems, it was acknowledged that depositions occurred both on land and at sea. But preference was clearly given to the on-shore depositions and problems.

Overall, the CLRTAP regime meshed well with Norwegian interests in many respects.[19] First, at the outset, the LRTAP framework convention must be characterised as an open and potentially dynamic regime (see Gehring, 1994). The framework character of the Convention, with few and open provisions, provided ample room for policy entrepreneurs to strengthen and develop the regime further. This room for manoeuvring has clearly been utilised by Norway and the other Nordic countries. Nordic negotiators and scientists have over time acquired a strong standing within the various CLRTAP bodies. For example, in the late 1990s a Norwegian was chairing the Executive Body, while both the chairman of the Working Group on Strategies and the head of the CLRTAP Secretariat were Swedes. Furthermore, the weight given to enhancing scientific knowledge in the Convention necessitated the establishment of a substantial 'complex' of scientific and technological working groups. Given Norway's interests and substantial scientific/technical competence in this issue area, Norway acquired both formal and informal leadership roles within this institutional complex.

In terms of decision rules, CLRTAP is built on a flexible consensus, hence reluctant countries have not blocked the adoption of protocols completely; they have simply not signed them. For instance, although the UK, as a great emitter of SO_2, opposed the 1985 Sulphur Protocol, it could not block its passage. With the benefit of hindsight, the adoption of this first protocol must be clearly be seen as a crucial victory for Norwegian interests because it set in motion a process of emission reductions that benefited the Norwegian environment. The regional focus, which limited the number of participants compared to a global effort,[20] also made it possible for an alliance of the Nordic countries together with countries such as Germany and the Netherlands to significantly influence decision-making and outcomes. It can be noted that the US and Canada are members of the UNECE and have also participated in CLRTAP, but neither of these countries have been central actors within the regime.

With regard to verification of compliance, at least initially, CLRTAP must be characterised as a fairly standard international environmental regime. Verification is mainly based on national reports. Compliance is reviewed at the yearly meetings of the Executive Body. The Secretariat plays an important role in this connection by preparing annual reviews and four-year major reviews. The EMEP monitoring network provides a certain capacity for independent verification at the overall regional level, but this capacity has not been used actively to check national compliance within the regime. The 1994 Second Sulphur Protocol introduced a specific Implementation Committee, which was established in 1997 and concentrates on reviewing reporting procedures and practices. Part of this work has consisted of publishing overview tables of reporting 'scores'. However, more recently a significant part of the work has concentrated on parties who are in non-compliance with their VOC commitments (Wettestad, 2003). As it is only in more recent years that CLRTAP has developed some 'teeth' with regard to the verification of compliance, this particular feature of CLRTAP has been of minimal benefit to Norwegian international goal attainment. On the other hand, the increasing and critical attention given to the VOC non-compliance of Norway and other states has probably contributed to the recent domestic strengthening of VOC policy (further described below).

In sum, the very receptivity of the core regime and the weight given to knowledge improvement were both regime characteristics that facilitated Norwegian goal achievement. These features provided room for Norwegian scientific and political entrepreneurs. Consensual decision rules, which did not allow single holdouts to block decisions, and the regional context, where a Nordic/North European alliance was a political heavy-weight, must also be counted on the positive side. However, CLRTAP's compliance review system was until recently fairly standard rather than particularly strong, and hence no important asset for Norway. Did the regime's strength then contribute to emission reductions abroad? The answer is positive, but the effects are diffuse and gradual. In themselves, neither the 1985 Sulphur Protocol nor the 1988 NO_x Protocol were very ambitious instruments. Moreover, in order to fully understand emission reductions abroad, the issue of linked regimes and institutions needs to be taken into consideration. In this case, this is particularly related to developing EU policies in this issue area (Wettestad, 2002). With regard to the significant reduction of European SO_2 emissions over time, it is for instance clear that the EU 1988 Large Combustion Plant Directive has contributed positively.[21]

Turning then to the domestic dimension and level of consensus on targets and coordination procedures, in the first decade of international cooperation, the Ministry of the Environment was pretty much alone and the clear leader in the domestic preparation processes.[22] The first SO_2 process took place without conflict, primarily because by 1984-85 Norway had already achieved most of its 30 per cent reduction target, and there was hence little left to fight about (Laugen, 1995, p.41). The level of inter-ministerial agreement seems to have been high also in the following NO_x preparatory process. There was overall agreement on the notion that NO_x emissions would be harder to reduce for Norway, and hence stabilisation of emissions was the most feasible position to be taken by Norway. However, the process was engulfed by the green tide that was affecting both the general public and the political sector alike. Norwegian negotiators were dismayed when Norwegian environment minister Sissel Rønbeck signed a political declaration to cut emissions by 30 per cent by 1998. The move can be seen to be a result of 'green competition' between Rønbeck and her Swedish counterpart Birgitta Dahl. In addition to this competition between progressive nations at the international level, fears of parliamentary whipping must also be brought in to understand why the Minister of the Environment signed such an ambitious declaration.[23]

Other ministries, and especially the Ministry of Finance, became more influential from the early 1990s process on VOCs.[24] The Ministry of Finance agreed whole-heartedly with the overall goal of bringing down European emissions and realised, along with other central players, that cutting NO_x and VOC emissions would entail real costs for Norway as a major oil and gas producer and a maritime nation. Hence, the Ministry of Finance was instrumental in bringing in the goal of cost effectiveness. But it was allegedly more sceptical with regard to the concept of differentiated agreements based on critical loads.

In the negotiations leading up to the 1999 Gothenburg Protocol, more ministries were involved than ever. In addition to the ministries of the environment, finance, and trade and industry, the ministries of oil and energy, transport, and agriculture, as well as the Prime Minister's Office, all participated – albeit only in one meeting each for the last three.[25] This much more inclusive process can probably be seen as a result of learning, on behalf of both the government and societal sectors. Moreover, it is a reflection of the unprecedented inclusive multi-pollutant approach, automatically affecting a number of societal sectors. According to participants in the process, the really tough political battles were fought at the domestic level, and the subsequent international negotiations appeared almost simple and friendly in comparison. Overall, domestic consensus must still be characterised as quite strong, and this has been a clear strength in Norway's international work in this issue area.

Explaining Domestic Problems: A Weak International Push and Domestic Disintegration?

A CLRTAP lamb and an EU tiger? Can international commitments and their strength shed any light on domestic performance? In the case of CLRTAP, only moderately it would seem. In the case of SO_2, as further substantiated by Laugen (1995), most of the 30 per cent emission reduction was achieved before the first Sulphur Protocol was signed in 1985. The further reductions in the wake of the 1985 Protocol have taken place much according to domestic processes only remotely related to the international process (the SO_2, NO_x and VOC domestic stories will all be further elaborated below).

In the case of NO_x, there is little to indicate that the 1988 Protocol's stabilisation commitment was much of a driving force in the domestic process. This has, of course, something to do with the fact that it was not a very ambitious target, and its achievement seemed to necessitate almost no new domestic action. On the other hand, the 1988 Declaration's much more ambitious 30 per cent target clearly required new and additional efforts. The Ministry of the Environment made a certain effort in the early 1990s to establish the 30 per cent reduction target as a driving political force, but had to give in by the mid 1990s due to opposition from other ministries. This was at least partly related to the fact that the 30 per cent target was a political, non-binding target. The VOC story is somewhat similar. Environmental authorities tried repeatedly to use the 30 per cent Protocol commitment as an instrument to establish effective domestic policies, but did not enjoy success until quite recently. This shows that even binding CLRTAP commitments have carried limited political weight so far, especially in cases where they have come up against the powerful Norwegian oil and gas interests.

So it may well be that the strength and political weight of international commitments carries most explanatory power in the case of the EU air pollution directives implemented through the European Economic Area (EEA) agreement. Norway was one of the central initiators of the EEA agreement in 1993, and although the country voted against becoming a full-fledged EU member in 1994, Norway has become subject to most of the EU air pollution policy and directives through the EEA agreement. As has been pointed out by Dahl (1999), Norway began to adapt to EU environmental policy as early as the late 1980s, and the process picked up speed after the formal establishment of the EEA agreement. Norway has generally shown a strong willingness to, at least *formally*, adapt to and follow-up EU directives. This is an expression of the high economic and political importance of the EU for Norway.

EU influence is relevant both in the case of VOCs and air quality. In terms of VOCs, for instance, it is clear that EU Directive 99/13 has spurred the Norwegian 2001 regulation of VOC emissions from solvent-using industries (PCA, 2001b). Moreover, when Norwegian authorities adopted air quality limits in 1997, an explicit and specific reference and link was made to two EU Directives.[26] But it is even more clear that the on-going revision of Norwegian air quality policy has been directly spurred by recent developments within EU. In the wake of the 1996 EU air quality framework directive, three daughter directives specifying air quality target values have been adopted. First, Directive 1999/30, which set standards for SO_2, NO_x, PM_{10}, and lead. Second, Directive 2000/69, which set standards for benzene and carbon monoxide. These two directives set initial limits from the time the directives entered into force, with a gradual tightening of the limits by 2005 and 2010. A third directive on ground-level ozone (Directive 2002/3) was adopted in February 2002.[27]

Domestic Disintegration: Horizontal as well as Vertical?

Institutions and legislation: a brief introductory overview. The Ministry of the Environment is responsible for coordinating policy formulation and implementation processes. The main legal basis is the 1981 Pollution Control Act, which replaced earlier separate air and water pollution legislation. The Act mainly covers pollution from stationary sources. Under the Ministry, there is one main pollution control directorate: the Pollution Control Authority, which is responsible for issuing permits under the Ministry of the Environment, monitoring programmes for air pollution, and setting emission standards for mobile sources. But the Ministry of Transport has the overall responsibility for transport and infrastructure policies. For ships, emission standards are set by the Maritime Directorate, which is subordinate to the Ministry of the Environment in these matters. With regard to the oil and gas sector, the Petroleum Act requires a plan for development and operation before a licensee/operator can develop a discovery. As part of this process, the operator must submit an environmental impact assessment. Apart from possible requirements imposed during consideration of this plan, offshore NO_x emissions have been unregulated. Offshore VOC permits are issued by the Pollution Control Authority.

In general, the Norwegian regulatory approach can on one hand be characterised as quite centralized, with the Pollution Control Authority regulating industrial emissions by issuing individual permits.[28] This has been supplemented by a financial assistance program initiated in 1975 (Ministry of the Environment, 1975) (see below). On the other hand, there is also a tradition of close cooperation between the governmental authorities and their respective societal sectors.[29] The

Parliament has input to decision-making processes mainly through reports – produced either in connection with policy formulation processes or protocol ratification (e.g. the CLRTAP protocols) – and related debates.

Political dynamics and policy instruments. Let us then turn to dynamics of the domestic political situation, starting off with the SO_2 story and then turning to the NO_x and VOC stories in turn. The 'shining sulphur success' is due to a combination of both luck and successful policies (Laugen, 1995). Leaving aside the element of luck for now (which will be further elaborated below), Norway's success in terms of substantially reducing its SO_2 emissions can to some extent be attributed to the adoption of adequate policy instruments. Sulphur emissions in Norway stem from industry and mining activities. The process of cleaning up industry started in the mid 1970s, well before the LRTAP Convention was even adopted. A central element was regulation of industrial emissions by individual concessions, motivated by local air pollution problems. As indicated above, this was part of a general pollution clean-up programme – although the measures to mitigate water pollution received most of the public financial muscle.[30] In terms of (other) economic instruments, a tax on the maximum SO_2 content in oil was introduced in 1976, and substantially stepped up from 1988 onwards.[31] This has contributed moderately to the increased use of light fuel oils, with a more moderate SO_2 content.[32] As a more specific EEA follow-up, regulation of and an additional levy on the quality of fuels and auto-diesel were introduced in the beginning of 2000. According to Statistics Norway, this led to a swift reduction by the oil companies of the SO_2 content in auto-diesel fuel.[33]

The NO_x story is a much more complicated one. The main bulk of Norwegian NO_x emissions stems from mobile sources (road traffic and ships), and the second main contributor is the petroleum sector.[34] As indicated, Norway did comply with the main and binding 1988 Sofia Protocol target of stabilisation by 1994. The question is, however, to what extent this compliance was enabled by the adoption of adequate policy instruments. On one hand, vehicle emissions requirements have clearly played a role (private petrol-fueled cars in 1989 and diesel-fueled in 1990; trucks in 1992 and 1993). According to the Pollution Control Authority, the vehicle requirements have counteracted the increase in emissions which otherwise would have taken place in the first part of the 1990s due to increased traffic and increased oil and gas production.[35]

But this statement also provides a clue to understanding why, on the other hand, Norway has fallen far short of living up to the 1988 political pledge to reduce emissions by 30 per cent. The vehicle requirements have not been accompanied by other effective political instruments. This does not mean that no efforts have been made to come up with such instruments and policies. The Pollution Control Authority initiated work in 1988 on a specified implementation plan to achieve the 30 per cent target, and the plan was submitted to the Ministry of the Environment in the fall of 1991. The ministry was not interested in publicity around the plan at this stage, and key points were made public only through leaks. The report discussed a long list of possible measures and their cost efficiency. Several packages of measures, each leading to achievement of the 30 per cent goal, were introduced and discussed. This phase (1988-91) was mostly internal within the 'environment segment' – with processes within the ministry and between the ministry and the Pollution Control Authority. But the Pollution Control Authority also kept in contact with several technological experts, such as MARINTEK/SINTEF in Trondheim, with regard to reducing ship emissions and emissions from platforms. However, according to the Ministry of the Environment, the time had now come to embark on a broader consultative process, discussing the different measures and packages with potentially affected ministries and social sectors.

It should also be noted that the NO_x issue was discussed in connection with several other on-going policy processes. First, NO_x reduction effects were integrated (regarding car exhaust limits) in a VOC policy package prepared by the Pollution Control Authority. Moreover, the climate change issue had been put firmly on the agenda at this point in time, and NO_x effects were included in the policy exploring process related to this issue. This was related to the fact that societal activities such as transport and petroleum production were central emitters of both NO_x and greenhouse gases such as carbon dioxide (CO_2). Third, there was an on-going international process within the International Maritime Organization (IMO) focusing on emissions to air from shipping. In 1991, the Ministry of the Environment initiated a first, seemingly quite informal, consultation process with other potentially affected ministries.[36] The consultation process hence involved important potential implementing agencies like the ministries of transport and energy and the Directorate of Shipping and Navigation. There are reasons to assume that the response was rather lukewarm in the various agencies,[37] as the next step from the Ministry of the Environment was a request for an additional report from the Statistics Norway and the Pollution Control Authority on societal benefits from NO_x reductions.

An interministerial coordinating committee was then established, consisting of the ministries of the environment, finance, foreign affairs, agriculture, transport and communication, industry, and energy. However, the committee's mandate was not restricted to NO_x; the issue of greenhouse gas emissions was also included. The general background for this issue linkage probably had to do with these two problems stemming from much of the same societal activities, as indicated above. Although not explicitly pointed out, the hope was probably that such a broadening of the agenda could result in synergy effects and the adoption of measures which seen only as air pollution or climate change measures were not feasible.[38] This was no success, as a classic case of horizontal disintegration unfolded. As pointed out by Laugen (1995, p.50), 'The Ministry of Finance...resisted more government spending; the Ministry of Transport...resisted measures to reduce traffic; the Ministry of Energy...resisted measures towards oil-production, and the Directorate for Shipping...resisted regulation of coastal traffic'.

The committee work was finished in 1994, and the report contained few specific, new NO_x regulatory initiatives (Ministry of the Environment, 1994). One exception was the establishment of a temporary (1996-2000) programme to help finance NO_x-reducing motor technical ship measures.[39] This progress in the area of shipping occurred more in spite of than due to developments within IMO, however. IMO was making very slow progress in this field, and pending IMO clarification was used in White Paper no. 41 (Ministry of the Environment, 1994) as an argument for a go-slow approach in this issue area.[40] So the NO_x and climate change issues were formally de-linked again. As the green tide had waned both in the public opinion and political parties by the mid 1990s, and the issue of climate change dominated the environmental agenda, the NO_x issue and the 30 per cent target received decreasing attention.

Turning then to the outright VOC failure, this is partly explained by the NO_x story, as several of the policies which failed to be adopted in the NO_x context also would have reduced VOC emissions. This is especially relevant for the transport sector. But the bulk of VOC emissions in Norway (i.e. around 60 per cent) stem from the petroleum sector, particularly from evaporation during transfer of crude oil to tankers.[41] Road traffic and other mobile sources are the second largest sources of emissions and account for around 18 per cent.[42] As early as 1990, the Pollution Control Authority produced a list of eight possible measures to bring down emissions. But little action was taken. The Ministry of the Environment stated in 1994 that this was because primary attention was given to the work on the combined NO_x and climate change measures (NM Bulletin, 1994). At the same time, the Ministry of the Environment claimed that it had control over the VOC process (ibid.). In 1995, it sent a renewed request to the Pollution Control Authority for a list of VOC measures (PCA, 1997, p.8). During the fall of 1995, the Ministry of the Environment sent out ambiguous signals with regard to the VOC process. In October, a high-ranking official declared that the 30 per cent reduction would be met (NM Bulletin, 1995a). However, in December, Environment Minister Thorbjørn Berntsen expressed serious doubt about whether the 30 per cent target could be met. But some action would be taken, and specific reference was made to work in progress on establishing voluntary agreements with the oil industry on reduced emissions (NM Bulletin, 1995b, p.7). This work took place in a broad contact forum involving ministries, oil companies, the fishing industry, research institutes, and environmental NGOs. The forum was called MILJOSOK and was established by the Ministry of Energy in 1995 (MILJOSOK, 1996).

The Pollution Control Authority then came up with its report on possible VOC measures in 1997, recommending especially the re-designing of tanker ships in order to reduce VOC evaporation as a key measure. Negotiations were then started in the spring of 1998 between governmental agencies and 17 oil companies on a voluntary agreement on technology development which could recycle VOC evaporation at the tankers and hence bring down VOC emissions significantly. More specifically, the aim of the agreement was to apply Best Available Technology (BAT) on the twenty relevant ships by 2005. The estimated cost would be around NOK 2 billion and lead to emission reductions in the order of 70 per cent from each ship. These negotiations stalled in December 1999, when major US oil companies (with Esso in the forefront) quite surprisingly refused to support the deal as a matter of principle. Esso was possibly afraid of setting a dangerous precedent. This development then led the Pollution Control Authority in October 2000 to change tactics and instead order the companies to reduce emissions by 95 per cent by 2005 (*Aftenposten*, 2000). Responding to dissatisfaction among the oil companies, a high-ranking Pollution Control Authority official declared that Norway was 'in severe non-compliance with the CLRTAP agreement which had a 1999 deadline. Time has come for action'.[43] The Pollution Control Authority's decision was then immediately challenged by all the involved oil companies,[44] meaning that it was up to the Ministry of the Environment to make a final decision. The ministry opted for a somewhat modified version of the Pollution Control Authority's requirements in November 2001. The main change was that the companies were given somewhat more lenient start-up deadlines with regard to the installation of abatement technology.[45]

The oil industry responded in June 2002 by establishing an agreement on the reduction of VOC emissions. Fifteen recovery systems are to be installed. Presently two tankers are equipped with such facilities, while six are being built. 23 industrial actors have signed this agreement.[46]

Table 5.2 The role of crossfire summarised

	The role of CLRTAP	The role of the EU	The role of other regimes/inst.	Crossfire summary
Norwegian SO_2 policy and performance	Very important as arena for influencing others. Little responsibility for high-class performance	• Some EU/EEA directives have influenced performance recently • EU LCP Directive contributed to reducing European emissions	Insignificant	Uncomplicated and friendly
NO_x policy and performance	Important as arena. A moderate impact on moderate performance	EU fuel standards gradually more important. 1999/2000 Air quality directives also relevant	Slow development of IMO standards and little synergy with climate measures have not been helpful	Somewhat complicated
VOC policy and performance	Important as arena. Has inspired policy-making efforts, but far from enough to counter-act powerful domestic actors	1999 Directive has led to a 2001 regulation	VOC measures in the early 1990s put on hold as NO_x and climate measures were clarified	Initial complications; later uncomplicated

Implementation failure due to lack of policy integration and 'penetration' of target groups? As can be recalled, a main assumption in this project is that goal attainment partly depends on the degree to which governments are able to integrate environmental policy and influence target groups. To what extent do such perspectives throw further light on the NO_x and VOC implementation trouble and failure summed up above?

Turning first to the issue of sectoral integration, let us zoom in more closely on the transport sector. Back in 1992, the Ministry of Transport outlined the Norwegian road and road traffic plan in White Paper no. 34 (Ministry of Transport, 1992). This Paper devoted some 25 pages to environmental challenges, and a substantial part of this was devoted to air pollution problems and various scenarios (ibid., pp.40-65). This discussion was followed up in White Paper no. 32 (Ministry of Transport, 1995), which outlined the basis for transport policy. The paper contained specific sections on environmental challenges, transport and environment, and traffic and environmental challenges in urban areas. With regard to international commitments, it stated, for example, that negotiations on a new NO_x protocol that would impose stricter targets for long-range transboundary air pollution were taking place, and that this protocol would influence transport policies. In terms of more general follow-up procedures in the relevant ministries, according to the recently published OECD Environmental Performance Review report, the procedural work did not really get into gear until 1997.[47] Hence, at least in the form of rhetoric, the transport sector has devoted some attention to air pollution issues from the early 1990s on. However, procedural and institutional sectoral integration has developed more slowly. In the sparsely populated and rocky Norway, transport ministers gain much more support by building tunnels and better roads than by giving priority to public transportation and environment-friendly solutions. So sectoral integration seems to have been slow and has had little effect on political priorities in practice. Hence, this may shed some light on the problems in terms of coming up with effective NO_x policies.

Influencing target groups has to do with the type of policy instruments utilised. As described in the case stories above, a mixed set of instruments has been applied – including regulation (e.g. sulphur licenses, vehicle emissions requirements), economic instruments (e.g. sulphur content of oil tax, financial aid for ship measures), and voluntary agreements (mainly the VOC agreement). In terms of shedding light on implementation failure, in the case of NO_x, the OECD has pointed to the lack of regulation of the 'significant' operational emissions from offshore petroleum installations (OECD 2001, p.160; Wettestad, forthcoming 2004). In the case of VOCs, Norwegian authorities backed the wrong horse when they put all their efforts into the voluntary VOC agreement.[48] When this agreement failed in the last minute of negotiations, they were left out in the cold, with no solid back-up measures. Hence, considerable implementation failure was inevitable.

Other Influential Factors: Problem Types

Norway's success on the international stage is of course only moderately its own doing. In the early phase of the functioning of CLRTAP, Norway's activity and influence was at its peak – together with Sweden and other Nordic countries. After Waldsterben and Germany's crucially important policy conversion in the early 1980s, Germany moved into the driver's seat both within CLRTAP and the EU. When attention turned to NO_x and VOCs, Norwegian interests became more complicated and Norway moved into more of a middle position in the international negotiations.[49] In order to explain the substantial emission reductions having taken place in central countries such as the UK, Germany and France, in addition to EU policies, several purely domestic and non-environmental processes in these countries must clearly be drawn into the picture. It is sufficient to mention privatisation and energy-switching in the UK and 'Wall Fall' effects in Germany when restructuring and closure of industries in East Germany brought down emissions 'for free' (Wettestad, 1996).

With regard to the mixed domestic Norwegian performance, fundamental issue characteristics have only been hinted at so far. But such differences must be brought in much more heavily in order to more fully understand why Norwegian performance varies so much in terms of SO_2 compared with NO_x and VOCs. With regard to the SO_2 success, compared to big European emitters such as the UK and Germany, Norway has first of all not been faced with the challenge of a costly retrofitting of large combustion plants. Moreover, several beneficial societal trends and events have significantly contributed to decreasing emissions. For instance, the closure of the copper mines in Sulitjelma alone meant a 10-12 per cent decrease in emissions 1980-93 (Ministry of the Environment, 1995, p.286). In addition, factors such as mild winters in the 1980s, closure of factories, and energy switching motivated solely by economic considerations have contributed (Ibid., pp.286-8). Although the total picture (for instance presented by Ministry of the Environment, 1995) is certainly complex, the main message is clear: targeted air pollution measures introduced by the Norwegian government within the broader context of CLRTAP, contributed very little to bringing down SO_2 emissions. Other contextual factors were far more important.

In a domestic comparative perspective, the problems of coming up with effective NO_x policies clearly have something to do with the problem type being less benign than in the case of SO_2. Norway has a long coastline, with many fjords and a related significant need for ferry traffic and sea transport. Moreover, the oil and gas industry forms the core of the national economy, and introducing costly abatement measures in this sector means taking on powerful interest groups. In addition, developments and trends which helped bring down SO_2 emissions have not benefited this issue area in a similar manner. For instance, oil and gas production has only increased over time.[50] This effect is seen clearly in the case of VOC. Although emissions from mobile sources decreased by 29 per cent in the period 1990-99, 'a 54 per cent growth in emissions from oil and gas operations masked this progress', as pointed out by the OECD (OECD, 2001, p.142). According to the Ministry of the Environment (2002, p.111), the oil production in 1999 was in fact 140 per cent higher than expected in the prognoses produced in connection with negotiations on the VOC Protocol in the beginning of the 1990s. So the VOC failure shares some similarities with the NO_x story. In order to achieve significant VOC reductions, one needs to take on the economically and politically key oil and gas sector. However, as significant reductions could have been achieved by a relatively simple measure targeted at a small group of actors, the VOC failure is also very much a case of choosing the wrong policy instruments.

Summing Up and Looking Ahead: Can the Gap Between International and Domestic Performance be Closed?

Summing up, its position as a major net importer of transboundary air pollution with vulnerable soil characteristics has spurred Norway to seek strong international commitments, not least with regard to SO_2, in order to bring down transboundary flows of pollutants. Over time, this central goal has been supplemented by an ambition to establish differentiated, 'smart' agreements based on critical loads. In this respect, Norway's goal attainment must be characterized as high. So the international chapter of the story is a very successful one. However, international success has not been matched by equally successful domestic policies. True, the domestic SO_2 performance is a shining success story. But the NO_x performance is barely adequate in relation to initial CLRTAP commitments (and clearly inadequate in relation to the 30 per cent Declaration target), and the VOC performance is simply a fiasco. This picture is roughly mirrored in the case of air quality, which overall was inadequate except for reductions in SO_2. However, the main deadlines are here 2005 and 2010, so there is some time yet to make improvements.

How, then, do we account for such a big difference between international and national goal attainment? Norway and Sweden clearly played an important role in the process of establishing CLRTAP and, over time, developing the critical loads approach and differentiated agreements. Moreover, the receptive character of the regime, the weight given to knowledge improvement, and the room allowed for Nordic scientific and political entrepreneurs have all been positive factors for Norway. But although CLRTAP has been important as an arena for knowledge improvement and policy development, other factors and developments have probably been more important for the international policy progress and for bringing down European emissions. For instance, in terms of catalytic significance for the development of both CLRTAP and EU air pollution policy, the importance of the German about-turn from laggard to leader in the early 1980s can hardly be overrated (See e.g. Boehmer-Christiansen and Skea, 1991; Wettestad, 2001).

Accounting for the varying Norwegian goal achievement, there are of course a number of explanatory factors, combining fundamental problem types, societal trends and events, and governmental measures and policies. For instance, in order to understand the much better SO_2 performance than NO_x and VOC performance, there is a combination of a regulatory more benign SO_2 problem (e.g. mostly industrial point sources); more benign societal trends and events in the case of SO_2 (e.g. the closing down of mines and no effect of increasing oil and gas production on these particular emissions); and possibly also more effective SO_2 policy measures (e.g. an SO_2 tax substantially stepped up over time). More specifically, the moderate NO_x performance may be best understood as a result of sectoral resistance and hence inadequate sectoral policy integration. The VOC failure has been very much related to the choice of the wrong policy instrument (i.e. a voluntary agreement which failed to come into existence). However, the common underlying factor impeding the development of effective policies in the NO_x and VOC cases is clearly the growth of Norway as an oil and gas producer and exporter in the 1990s.

This project is particularly concerned with the possibility of the domestic performance being either enhanced or hindered by the activities of other international institutions than the core international institution. As indicated, the core institution in this connection is CLRTAP. Has the shining sulphur success been enhanced by other institutions than CLRTAP? As has been shown, CLRTAP did not contribute much to the Norwegian SO_2 performance, and other institutions such as the EU have not played much of a role in this context either (although the EU has contributed to bringing down European emissions).

With regard to the medium NO_x performance, the interplay between CLRTAP and other institutions has not been very synergistic. For instance, the slow development of IMO ship emissions rules has not made the task of curbing the emissions from this sector easier. Moreover, the potential synergy effects between NO_x and climate change measures have not materialised. On the mildly positive side, the NO_2 part of the follow-up of EU air quality directives can be noted. But as the EU directives are of a quite recent nature, their effect has just started.

In the case of VOCs, the linked NO_x and climate change process in the first part of the 1990s was used as an excuse for going slow in terms of initial VOC policy development. More recently, EU/EEA requirements have acted as a mild positive force in turning the trend of increasing emissions. Other than that, international institutions can neither be credited nor blamed for the lacklustre Norwegian performance in this issue area. In the case of air quality, the most relevant international institution has been the EU. The importance of the EU air quality policy development and directives for Norwegian goal formulation has been significant.

Looking ahead, both the 1999 Gothenburg Protocol and recent EU directives necessitate a renewed and strengthened link between external requirements and domestic performance in the Norwegian air pollution policy debate. It will be embarrassing for Norway to end up in 2010 as a relatively clean country due to the considerable efforts of others – and as one of the worst Western European non-compliers within the CLRTAP context. So what are the main prospects for Norwegian performance ahead? And how may the links to other institutions affect this performance? Based on the previous sections, the issues of NO_x, VOCs, and air quality are the most pertinent challenges.

Turning first to NO_x and possibly the most difficult challenge, the Pollution Control Authority has in recent years expressed optimism with regard to the prospects for bringing down emissions by 28 per cent in the decade ahead, as required by the Gothenburg Protocol. In terms of vehicle emissions, it is interesting to note that the Pollution Control Authority emphasizes EU requirements as a main, pushing factor. Hence, a 1999 report states that emissions from road traffic are expected to decrease by 50 per cent between 1998 and 2010 as a result of

Norway adopting stricter EU fuel and vehicle emission requirements (PCA, 1999a, p.8). The same report expects emissions from the shipping sector to decrease in response to new MARPOL/IMO motor requirements. However, in the more recent White Paper from the Ministry of the Environment (2002), the emissions from the shipping sector are pointed out as the most serious challenge (ibid., p.112). Another challenge is of course the emissions from the petroleum sector, and these emissions are expected to increase by more than 20 per cent (PCA, 1999a, p.8). But the installation of low-NO_x burners on North Sea oil rigs may soften this increase.[51] Overall, a number of possible measures have been identified in several reports from the Pollution Control Authority and Ministry of the Environment, but the final package has not so far been decided. It is interesting to note that White Paper no. 25 (Ministry of the Environment, 2002) also brings in the 1996 EU Integrated Pollution Prevention Control (IPPC) Directive as a central element in the further regulatory process (ibid., p.112).

As has been noted earlier, although VOC emissions are far higher than required by CLRTAP, these emissions have decreased somewhat in recent years. This development is expected to continue due to increased VOC recycling at an on-shore oil terminal, reduced evaporation during transfer of crude oil to tankers, and reduced emissions from traffic (ibid., p.111). It is clear that the tanker loading and storage requirements adopted by the Pollution Control Authority and Ministry of the Environment in 2000-2001 will bring down emissions considerably. But the OECD is not very optimistic with regard to compliance: 'as [the] Geneva Protocol target was so widely missed, and in view of the importance of the petroleum sector, it is not obvious how Norway could achieve this reduction through domestic measures' (OECD, 2001, p.161). However, according to the Ministry of the Environment, Norway will be in compliance with the CLRTAP commitments from 2005 on (Ministry of the Environment, 2002, p.111).

With regard to the air quality targets, as was noted earlier, the current Norwegian targets are generally stricter than those adopted by the EU. According to the Ministry of the Environment (2002), it will not be problematic to meet the SO_2 and benzene targets. As for particulate matter, stricter EU fuel standards will contribute positively, both by reducing transboundary import from the EU area and by tightening up domestic standards. But the White Paper warns that tightened or new measures are most probably needed in order to attain targets (ibid., p.114). With regard to NO_x and NO_2, there is of course a clear link to the success of the NO_x measures discussed above. If these measures are successful, this will help in the process of attaining some of the air quality targets also. However, it should be noted that the White Paper mentioned above also contains a similar warning for NO_2 as for particulate matter with regard to the chances for goal attainment in 2010 (ibid., p.114).

Winding up, there are signs of increasing governmental awareness of the challenges posed by recent CLRTAP and EU policies and the need to come up with more effective domestic policies. It is clear that the policy instruments are available; it is more a matter of finding the right package or mix in the differing issue areas. Perhaps the increasing interplay and complementarity of CLRTAP and EU policies witnessed in the recent years can provide the Ministry of the

Environment and Pollution Control Authority with the necessary authority to get adequate policy packages adopted and implemented in the decade ahead. It may hence be symptomatic that the Pollution Control Authority referred to the EU IPPC Directive as the central driving force when stricter offshore air emissions requirements were signalled in December 2002 (PCA, 2002b).

Notes

1 This chapter has benefited from input and comments from Anita Wahlstrøm, Eli M. Åsen (both Ministry of the Environment) and Erland Røsten (Norwegian Pollution Control Authority). Special thanks to Jan Thompson for providing valuable comments and Lynn P. Nygaard for helping me to improve the language and clarity of the chapter.

2 For information on all Protocols under CLRTAP, see United Nations Economic Commission for Europe, Environment and Human Settlements Division, www.unece.org/env/lrtap/.

3 The baseline here was 1987.

4 See the 1988-89 White Paper on environment and development (Ministry of the Environment, 1989, p.60). Here, 1986 was specified as the baseline.

5 A critical load can be defined as a quantitative estimate of an exposure to one or more pollutants below which significantly harmful effects on specified sensitive elements of the environment do not occur according to present knowledge.

6 For an overview of the VOC negotiations, see Gehring (1994).

7 The baseline was 1989.

8 The CLRTAP Oslo Protocol sets out individual and varying national reduction targets for the year 2000 for half of the countries, and additional 2005 and 2010 targets for the other half – with 1980 as base year. See www.unece.org/env/lrtap.

9 As the next regulatory steps within the CLRTAP context, negotiations on protocols on the emissions of heavy metal and persistent organic pollutants (POPs) were conducted in 1997-98. The 1998 Aarhus Protocol on heavy metals targeted three specific substances: cadmium, lead and mercury. Along with the other parties, Norway has agreed to reduce emissions of the three substances mentioned above below their levels in 1990 (or an alternative year between 1985 and 1995); however, without a specified deadline. The second 1998 Aarhus Protocol targeted POPs. The production and use of some POPs were banned outright (i.e. aldrin, chlordane, chlordecone, dieldrin, endrin, hexabromobiphenyl, mirex, and toxaphene); others were scheduled for elimination at a later stage (i.e. DDT, heptachlor, hexachlorobenzene, PCBs). Moreover, the protocol severely restricted the use of DDTs, HCH and PCBs. See YBICED 1999, pp.85-86. As the discussion of goals and follow-up concentrates on the substances with a long and incremental regulatory history, heavy metals and POPs are not included in the further discussion.

10 See United Nations Economic Commission for Europe, Environment and Human Settlements Division, www.unece.org/env/lrtap.

11 See summary and comparison with EU standards at the time in Ministry of the Environment, 1995, p.249. For both NO_x and particulate matter, Pollution Control Authority's standards were three times as stringent as the EU standards.

12 For instance, in a crude comparative perspective, the 1992 particulate matter limit of 70 ug/m3 was reduced to 50 ug/m3 in 1997. However, it should be noted that a more

sophisticated comparison of stringency also needs to take into account parameters such as number of exceedance days allowed.

13 For instance, with regard to NO_2, the hourly 2005 limit was set at 300 ug/m3.

14 Except for SO_2, which has a 2005 deadline. Moreover, in comparison with the binding 2005 NO_2 limit, the 2010 goal is that the hourly concentration of NO_2 shall not exceed 150 ug/m3 more than 8 hours per year. See Ministry of the Environment, 1999, p.107.

15 See PCA (2001a). Note, however, that the strategic national goals from 1998 remain the same.

16 See the presentation of IIASA's assessments in EU Commission (1997).

17 Statistics Norway, 'Emissions to Air', 1999 (www.ssb.no). According to the Ministry of the Environment, (2000, p.98) the increase in emissions was 15 per cent.

18 Except for some problems in the city of Sarpsborg caused by local industry. See Ministry of the Environment (2002, p.115).

19 For a more comprehensive discussion of CLRTAP's institutional design, see Wettestad (1999), Chapter 4.

20 Only 24 countries participated in the negotiations leading up to the 1985 Protocol. The current number of parties is 48. See www.unece.org/env/lrtap/.

21 For an overview of the 1988 Large Combustion Plant Directive and effects in the UK, see Haigh (ed., 2002, section 6.10).

22 According to Harald Dovland and Eli M. Aasen in the Ministry of the Environment, the ministry took care of the SO_2 and NO_x negotiatons alone. Communication with Dovland and Aasen, January and February 2002.

23 See also Laugen (1995, p.42).

24 According to Eli M. Aasen, the Norwegian VOC delegation consisted of the ministries of environment, finance, and trade and industry. Communication with Aasen, Ministry of the Environment, February 2002.

25 Communication with Eli M. Aasen, Ministry of the Environment, February 2002.

26 These Directives were 80/779 and 85/203. See Ministry of the Environment (1997).

27 See Haigh (ed., 2002). For an overview of recent developments in EU air pollution policy, see Wettestad (2002).

28 This approach makes sense, as there are few large combustion facilities, and point-source standards such as those applicable to large combustion facilities in the relevant EU directives would not apply in Norway's situation. See OECD (1993, p.68).

29 See for instance Laugen (1995, p.69); OECD (1993, p.88).

30 In the issue area of air pollution, governmental spending has mainly gone to liming measures. Annual spending has been around NOK 100 million. Communication with Eli M. Aasen and Steinar Hermansen, both from the Ministry of the Environment, February 2002.

31 See Statistics Norway, Energy Statistics 1998, section 4.1, www.ssb.no.

32 See Statistics Norway, www.ssb.no, sulphur dioxide figures from 2000.

33 Ibid.

34 In the late 1980s, mobile sources accounted for around 70 per cent of the emissions. A decade later this sector's share has increased a little bit. Road traffic and ships initially contributed roughly equally; the trend here is an increasing share for ships. The petroleum sector accounted for around 13 per cent of the emissions. This sector's share of the emissions is currently around 15 per cent. See PCA website at www.sft.no.

35 See PCA (1999, p.7). According to this report, the main reason for decreasing NO_x emissions in Norway from 1987-92 was economic depression.

36 According to *NM Bulletin* (1992), the Ministry of the Environment spokesperson stated that recent months had been spent on 'several informal contacts' with relevant ministries.

37 According to *NM Bulletin* (1993), the initial ideas from the PCA created 'sharp reactions' and 'fury' in the potentially affected ministries.

38 For instance, the central Ministry of the Environment White Paper no.41 (1994) did not explicitly address this issue. However, the sectoral analyses in Chapter 8 in this Paper showed that CO_2 and NO_x emissions mainly stem from the same activities and hence, more implicitly, a case was made for an integrated approach.

39 This was financed over the budget of the Industry and Trade Ministry. Assistance to 45 ships was granted in the period 1996–2000, contributing to an annual decrease of about 1400 tons in Norway's NO_x emissions. See Ministry of the Environment (2000, p.100).

40 See Ministry of the Environment (1994, pp.141-2). The need for an international solution through IMO partly stemmed from the fact that significant NO_x emissions in Norwegian waters stemmed from foreign vessels.

41 The estimates of the VOC emissions related to this activity were revised upwards by 24 per cent in 1997. See *Acid News* (1997).

42 PCA web site (www.sft.no), December 2002.

43 PCA departmental director Marie Nordbye in *Aftenposten* (2000). Author's translation.

44 I.e. Statoil, Norsk Hydro, Esso Norway and Shell Norway.

45 For instance, the PCA had initially required a 30 per cent installation of VOC-reducing technology by the end of 2001. On the basis of consultations with the companies, the PCA recommended to the Ministry of the Environment that the 2001 deadline be cancelled and a 40 per cent requirement by 2003 be adopted instead. This was then adopted by the Ministry of the Environment. See Ministry of the Environment (2001a).

46 In addition to Statoil, Norsk Hydro, Esso Norway and Shell Norway, the following companies participate in this collaboration: Petoro, Fortum, TotalFinaElf, Norsk agip, BP, Dong, Idemitsu, RWE-DEA, ChevronTexaco, Paladin, Gaz de France, Enterprise Oil, DNO, Conoco, and Amerada Hess. See Shell website, News, 25 June 2002 (www.shell.com).

47 See OECD (2001, p.98). In 1997, eight environmental priority areas were identified, and all sectoral administrations were to establish environmental action plans and submit annual progress reports.

48 The background for this is probably a combination of a general desire to please the industry and an increasing interest all over Europe in the use of voluntary agreements as an environmental policy instrument.

49 However, as noted earlier, Norwegian and Swedish scientists and negotiators have continued to play important roles in the development of the regime, not least in the development and application of the critical loads approach.

50 With regard to the most recent decade, oil production in 1990 was around 150 billion scm oe (bn scm), and increased up to 230/240 bn scm in 2001; averaging 3.1 million barrels a day. With regard to natural gas, production in 1990 was around 30 billion scm, and increased up to around 60 billion scm in 2001 (and is expected to further increase to around 100 bn scm in 2010). See Ministry of Petroleum and Energy (2002, p.9).

51 PCA (1999b). For more on the issue of low-NO_x burners, see Wettestad (forthcoming 2004).

References

Acid News (1997), 'Now in Jeopardy', no. 1, April, p. 5.

Aftenposten (2000), 'Enkle grep kan minske utslipp i Nordsjøen', 14 December.

Boehmer-Christiansen, S. and Skea, J. (1991), *Acid Politics: Environmental and Energy Policies in Britain and Germany*, Belhaven Press, London.

Chossudovsky, E. (1989), *East-West Diplomacy for Environment in the United Nations*, UNITAR, New York.

Dagsavisen (2002), 'Har ikke luftplan', January 13.

EU Commission (1997), Communication from the EU Commission on 'A European Strategy to Combat Acidification', reprinted as a supplement to *Europe Environment*, April 8, p. 497.

Dahl, A. (1999), 'Miljøpolitikk – full tilpasning uten politisk debatt', in D.H. Claes and B.S. Tranøy (eds), *Utenfor, annerledes og suveren? Norge under EØS-avtalen?*, Fagbok-laget, Oslo.

Gehring, T. (1994), *Dynamic International Regimes – Institutions for International Environmental Governance*, Peter Lang, Berlin.

Haigh, N. (ed., 2002), *Manual of Environmental Policy: The EC and Britain*, Longman/IEEP, London.

Laugen, T. (1995), *Compliance with International Environmental Agreements – Norway and the Acid Rain Convention*, report R:003-1995, The Fridtjof Nansen Institute, Lysaker.

MILJOSOK (1996), *Oljeindustrien tar ansvar*, report from the MILJOSOK Steering Group, 13 December.

Ministry of the Environment (1975), *Measures Against Pollution* (Tiltak mot forurensning), Report No. 44 to the Storting (1975-76), Ministry of the Environment, Oslo.

Ministry of the Environment (1984), *Measures Against Pollution* (Tiltak mot forurensning), Report No. 51 to the Storting (1984-85), Ministry of the Environment, Oslo.

Ministry of the Environment (1989), *Environment and Development. Programme for Norway's Follow-up of the Report of the World Commission on Environment and Development* (Miljø og utvikling. Norges oppfølging av Verdenskommisjonens rapport), Report No. 46 to the Storting (1988-89), Ministry of the Environment, Oslo.

Ministry of the Environment (1994), *On Norwegian Climate Change and NO_x Policy* (Om norsk politikk mot klimaendringer og utslipp av nitrogenoksider (NO_x), Report No. 41 to the Storting (1994-95), Ministry of the Environment, Oslo.

Ministry of the Environment (1995), *Types of Measures in Environmental Policy-making* (Virkemidler i miljøpolitikken), NOU report No. 1995:4, Ministry of the Environment, Oslo.

Ministry of the Environment (1997), 'The 1997 Regulation on Limit Values for Local Air Pollution and Noise, Chapter II (Limit values)', retreived from http://odin.dep.no.

Ministry of the Environment (1999), *The Government's Environmental Policy and the Condition of the Norwegian Environment* (Regjeringens miljøvernpolitikk og rikets miljøtilstand), Report No. 8 to the Storting (1999-2000), Ministry of the Environment, Oslo.

Ministry of the Environment (2000), *The Government's Environmental Policy and the Condition of the Norwegian Environment* (Regjeringens miljøvernpolitikk og rikets miljøtilstand), Report No. 24 to the Storting (2000-2001), Ministry of the Environment, Oslo.

Ministry of the Environment (2001a), *Requirements for Reducing Off-shore Oil Evaporation (VOC)* (Pålegg om reduksjon av oljedamp (VOC) på sokkelen), 13 November, Ministry of the Environment, Oslo.

Ministry of the Environment (2001b), 'Behind Acid Clouds' (Bak sure skyer), information brochure on acid rain, Ministry of the Environment, Oslo.

Ministry of the Environment (2002), *The Government's Environmental Policy and the Condition of the Norwegian Environment* (Regjeringens miljøvernpolitikk og rikets miljøtilstand), Report No. 25 to the Storting (2002-2003), Ministry of the Environment, Oslo.

Ministry of Petroleum and Energy, (2002), *Environment 2002 – The Norwegian Petroleum Sector*, Ministry of Petroleum and Energy, Oslo.

Ministry of Transport (1992), *Norwegian Road and Road Transport Plan 1994-97* (Norsk vei og transportplan 1994-97), Report No. 34 to the Storting (1992-93), Ministry of Transport, Oslo.

Ministry of Transport (1995), *The Basis for Transport Policy* (Grunnlaget for transportpolitikken), Report No. 32 to the Storting (1995-96), Ministry of Transport, Oslo.

NM Bulletin (1992), 'Tidspress for NOx-reduksjon', no. 15, 28 August, pp. 1-6.

NM Bulletin (1993), 'Regjeringen forlater NOx-mål', no. 14, 13 August, p. 1.

NM Bulletin (1994), 'VOC-plan utsatt – mål i fare?', no. 4, 25 February, p. 4.

NM Bulletin (1995a), 'VOC-utslipp gir alvorlig helsefare i Drammen', 20 October, p. 5.

NM Bulletin (1995b), 'Skrinlegger nok et miljømål', 15 December, p. 7.

OECD (1993), *Norwegian Performance Review*, OECD, Paris.

OECD (2001), *Environmental Performance Reviews – Norway*, OECD, Paris.

PCA, Norwegian Pollution Control Authority (Statens forurensningstilsyn, SFT) (1997), *Measures for Reducing NMVOC Emissions in Norway* (Tiltak for reduksjon av NMVOC-utslipp i Norge), Report No. 1997:11, Pollution Control Authority, Oslo.

PCA, Norwegian Pollution Control Authority (Statens forurensningstilsyn, SFT) (1999a), *Reducing NO_x Emissions in Norway* (Reduksjon av NO_x utslipp i Norge), Report No. 1999:13, Pollution Control Authority, Oslo.

PCA, Norwegian Pollution Control Authority (Statens forurensningstilsyn, SFT) (1999b), *How Much Will the New Gothenburg Protocol Cost Norway?* (Hvor mye vil Gøteborg-protokollen koste Norge?), note dated 9 December, Pollution Control Authority, Oslo.

PCA, Norwegian Pollution Control Authority (Statens forurensningstilsyn, SFT) (2001a), 'New and Stricter Air Quality Standards' (Nye og strengere krav til luftkvalitet), Pollution Control Authority Press Release, 10 October.

PCA, Norwegian Pollution Control Authority (Statens forurensningstilsyn, SFT) (2001b), 'Regulation of VOC Emissions from Solvent-using Industries', (2001-10-01 no. 1139).

PCA, Norwegian Pollution Control Authority (Statens forurensningstilsyn, SFT) (2002a), 'Environmental Conditions in Norway: Local Air Pollution' (Norsk miljøtilstand: lokal luftkvalitet), retreived in 2002 from the Pollution Control Authority web site www.sft.no.

PCA, Norwegian Pollution Control Authority (Statens forurensningstilsyn, SFT) (2002b), 'Stricter Norwegian Offshore Air Requirements to Come' (Strengere luft-krav på norsk sokkel), Pollution Control Authority Press Release, 5 December.

PCA, Norwegian Pollution Control Authority (Statens forurensningstilsyn, SFT) (2003), The Oslo Air Steadily Less Polluted (Oslo-lufta stadig mindre forurenset), Pollution Control Authority Press Release, 21 February.

Stenstadvold, M. (1991), 'The Evolution of Cooperation: A Case Study of the NO_x Protocol (in Norwegian), unpublished thesis, University of Oslo.

Wettestad, J. (1996), *Acid Lessons? Assessing and Explaining LRTAP Implementation and Effectiveness*, WP-96-18 March, IIASA Working Paper. A revised version was published as Wettestad, J. (1997), 'Acid Lessons? Assessing and Explaining LRTAP Implementation and Effectiveness', *Global Environmental Change*, vol. 7, no. 3, pp. 235-49.

Wettestad, J. (1998), 'Participation in NO_x Policy-making and Implementation in the Netherlands, UK and Norway: Different Approaches, but Similar Results?, in D.G. Victor, K. Raustiala and E.B. Skolnikoff (eds), *The Implementation and Effectiveness of International Environmental Commitments*, MIT Press, Cambridge MA, pp. 381-431.

Wettestad, J. (1999), *Designing Effective Environmental Regimes – The Key Conditions*, Edward Elgar, Cheltenham.

Wettestad, J. (2001), 'The Convention on Long-range Transboundary Air Pollution (CLRTAP)', in E.L. Miles, A. Underdal, S. Andresen, J. Wettestad, J.B. Skjærseth and E.M. Carlin, *Environmental Regime Effectiveness: Confronting Theory with Evidence*, MIT Press, Cambridge MA and London, pp. 197-223.

Wettestad, J. (2002), *Clearing the Air – European Advances in Tackling Acid Rain and Atmospheric Pollution*, Ashgate, Aldershot.

Wettestad, J. (2003), *Enhancing Climate Compliance: What are the Lessons to Learn from Environmental Regimes and the EU?*, FNI-Report 2/2003, The Fridtjof Nansen Institute, Lysaker.

Wettestad, J. (forthcoming 2004), Offshore Air Pollution and Technological Fixes: How a Greater Need for Norwegian than UK Action Finally Won Through, forthcoming FNI-Report, The Fridtjof Nansen Institute, Lysaker.

YBICED (1999), *Yearbook of International Co-operation on Environment and Development 1999/2000*, The Fridtjof Nansen Institute/Earthscan Publications Ltd., London.

Chapter 6

Marine Pollution: International Ambition, Domestic Resistance

Jon Birger Skjærseth

Introduction

Discharges of hazardous substances and nutrients (phosphorus and nitrogen) to the marine environment are transported by currents from one country to another. The counter clockwise direction of the North Sea currents places Norway as a net importer of marine pollution.[1] Not surprisingly, Norway has pushed for stringent international obligations in order to affect the behaviour of other North Sea states, notably the pollution exporters. Since Norway took the initiative to the first international convention on marine pollution (the Oslo Convention on dumping) in the North-East Atlantic in 1972, Norway has pushed for stringent international regulation of nutrients and particularly hazardous substances. As a result of Norway's efforts, and the efforts of other interested states, the North Sea and the North-East Atlantic marine environment are now addressed internationally through international treaties, political declarations and EU directives, regulations and decisions.

While Norway was one of the states that pushed hardest to get this international regulation in place, it faces significant challenges in practising at home what it preaches abroad. This is most clearly illustrated in the case of nitrogen: In 1988 and 1989, Norway was hard hit by toxic algae blooms along the Norwegian coast. These blooms were linked to emissions of nutrients, and Norway urged all North Sea states to strictly follow the international obligations adopted, i.e. a 50 per cent reduction of nutrients to sensitive areas between 1985 and 1995. However, Norway has not even come close to complying with the international obligation on nitrogen.

The aim of this chapter is to explore this gap between Norway's international and domestic goal attainment. The magnitude of this gap is discussed before exploring how it may have occurred. The chapter divides the time period under consideration into two periods: phase one, which is dominated by efforts to influence regime commitments at the international level; and phase two, which is dominated by implementation at the domestic level. Explanation is sought primarily in the role of institutions: domestic institutions, the core regime, and linked institutions. It then considers an alternative approach by looking at how problem type in terms of cost-benefit distribution and configuration of actor interests may have affected the outcome. It argues that the gap between international and domestic goal attainment has been caused by a combination of ambitious international obligations and domestic implementation problems.

Norway's Goal Attainment at the International and Domestic Levels

Growing Domestic and International Regulation of the Marine Environment

Norwegian legislation on water pollution may be traced back to the Water Pollution Act (WPA) of 1970. This act formed the basis for regulation directed at municipal wastewater, agriculture and industry – the three main sources of marine pollution. Sea-dumping of industrial waste and sewage sludge has never been viewed as a viable disposal option in Norway.[2] The WPA was later replaced by the Pollution Control Act (PCA) of 1981, covering air, water and ground. Up to the second half of the 1980s, water pollution legislation and programs on hazardous substances and nutrients were mainly motivated by domestic concerns. The WPA was adopted before any international convention on the North Sea or on the wider North-East Atlantic area.

Domestic vulnerability due to domestic causes and sources was the major motivating factor behind the comprehensive first-generation action program launched by the government in the mid 1970s (Ministry of the Environment, 1975). The first 1974-75 White Paper on water pollution commences by mapping the state of the environment both inland and along the coast. The government expected that clean-up of present and future water pollution problems would continue to the early 1990s, and investments were expected to decrease from the late 1980s. Norway had ratified the two main conventions on the marine environment in the North-East Atlantic: the 1972 Oslo Convention and the 1974 Paris Convention. Signed by all 13 West European maritime states, the Oslo Convention regulates dumping and incineration at sea along the entire North-East Atlantic up to the North Pole. The Convention for the Prevention of Marine Pollution from Land-based Sources – the Paris Convention – was signed in Paris two years later by roughly the same states as the Oslo Convention. The Oslo and Paris Conventions were supported by a joint secretariat, executive commissions (the Oslo and Paris Commissions) and several standing and ad hoc scientific/technical bodies.[3] The Paris Convention allowed the European Community to join as a contracting member, and water policy was the first sub-sector developed under EU environmental policy. The Paris Convention was considered by Norway to 'set an external framework for the measures needed to combat pollution from land-based sources' (Ministry of the Environment, 1975, p.10).

In the mid 1980s, the government produced a new White Paper on water pollution in which only one page was devoted to international cooperation (Ministry of the Environment, 1985). However, Norway stressed that international cooperation is crucial to combating marine pollution, and the work within the Oslo and Paris Commissions was perceived as progressing 'significantly slower than desired' (Ministry of the Environment, 1985, p.5). The domestic measures adopted between 1975 and 1985 were described as partly successful. Industry had been most successful in reducing hazardous substances, while local eutrophication problems were reduced as a result of investments in municipal wastewater treatment plants. Measures in the agricultural sector were perceived as unsatisfactory since they had been taken on point sources only (silos, manure). Still, the 1985 White Paper leaves an overall impression of satisfaction with the water-related environmental situation.

Germany shared the Norwegian dissatisfaction with existing international treaties and obligations. This dissatisfaction was rooted in growing indications in the early 1980s that specific parts of the North Sea were becoming severely polluted (Ehlers, 1990, p.4). Against this backdrop, Germany took the initiative to arrange the first International North Sea Conference (INSC) in Bremen in 1984. The Bremen Conference was followed by conferences in London in 1987, The Hague in 1990, Esbjerg (Denmark) in 1995 and, most recently, Bergen (Norway) in 2002. Conference participants comprised the eight North Sea coastal states and the EU and thus represented a sub-set of the original Oslo and Paris Conventions parties.[4]

The 1987 London Declaration represented a turning point in its ambition to phase out dumping of industrial waste and incineration at sea, to reduce inputs of nutrients to sensitive areas by 50 per cent between 1985 and 1995, and to reduce total inputs of hazardous substances reaching the aquatic environment by in the order of 50 per cent within the same time frame. The 1990 Hague Declaration clarified and strengthened the London Declaration, particularly concerning land-based sources. A list of 36 hazardous substances was adopted to reduce emissions of such substances by 50 per cent between 1985 and 1995. Targets aimed at a 70 per cent reduction of land-based and atmospheric inputs were adopted for the most dangerous substances – dioxines, cadmium, mercury and lead. The 1995 Esbjerg Declaration stands as a symbol of the significant gains made in ambitions on hazardous substances from the 1970s to the present. In the mid 1980s, only a few hazardous substances were under international regulation and even fewer were made subject to elimination. Ten years later, the North Sea Ministers agreed to prevent the pollution of the North Sea by phasing out all hazardous substances within 25 years:

> ... by continuously reducing discharges, emissions and losses of hazardous substances thereby moving towards the target of their cessation within one generation (25 years) with the ultimate aim of concentrations in the environment near background values for naturally occurring substances and close to zero concentrations for man-made synthetic substances. (Ministry of Environment and Energy, Denmark, 1995, p.18)

The Oslo and Paris Commissions together with the EU took significant steps in the same direction in the wake of the 1987 North Sea conference.[5] In 1992, the Oslo and Paris Conventions were brought together to form a single legal instrument for the protection of the North-East Atlantic and became known as the OSPAR Convention.

Norwegian domestic goals on water and marine pollution changed dramatically in line with the international development. Norway became increasingly aware of its role as a 'victim' of transnational pollution. We may note the opening speech made by the Norwegian Minister of the Environment at the 1987 North Sea Conference: 'Ocean currents carry pollutants, especially from the southern part of the North Sea into Norwegian waters. This gives us grounds for *deep concern*' (Department of the Environment, UK, 1987, p.1, emphasis added). In the 1988-89 White Paper prepared by the Ministry of the Environment, emissions of nutrients, mainly from agriculture and municipal wastewater, were to be reduced by 50 per cent between 1985 and 1995 in sensitive areas.[6] Moreover, discharges of hazardous substances, mainly from industry, were to be reduced in the order of 70 per cent between 1985 and 1995. It also called for the cessation of dumping and incineration at sea as soon as possible. In short, Norway had adopted all the joint international commitments on hazardous substances and nutrients as national targets.[7]

Norway's Goal Attainment

Norwegian water quality is affected by the emissions of domestic as well as foreign target groups. Emissions of nutrients and hazardous substances from domestic industry, agriculture, aquaculture and municipal wastewater cause water pollution problems in lakes and rivers and along the Norwegian coastline. Emissions from other North Sea states are also significant for the quality of Norwegian marine areas, where Norway has strong economic interests in fishing and aquaculture.

Internationally, Norway has generally pushed for stringent international commitments to counter long-range marine pollution. In the North Sea cooperation, the dispute between those states favouring Environmental Quality Objectives (EQOs) and those preferring Uniform Emission Standards (UES) hampered any progress on regulation until the late 1980s. Crudely put, UES advocates – including Norway – emphasized that discharges of substances known to be toxic, persistent and bioaccumulative should be limited as far as possible at source in line with the precautionary principle, whereas the EQO defenders – led by the UK – maintained that standards set should be determined by observable negative effects in the marine environment. This controversy was linked to both varying material interests and different norms concerning the basic philosophy of pollution control. The UES/EQO dispute led to cumulative conflict by pitting the same actors against each other on different substances (Skjærseth, 1999).

In the case of dumping at sea, the main target groups were located abroad, since Norway did not dump industrial waste, or sewage sludge. As noted, Norway took the initiative to establish the Oslo Convention in 1972, and since has, in alliance with the Nordic states, worked for a ban on dumping in the North-East Atlantic. The Nordic countries argued normatively in the Oslo Commission by stressing that using the sea as 'a trash can' was inherently wrong, in addition to leading to negative environmental consequences. From 1987 to 1990, the International North Sea Conferences as well as the Oslo Commission took various steps aimed at phasing out dumping completely. In 1990, the UK was the only North Sea state continuing to dump industrial waste. From 1990 to 1993, the UK reduced the amount of deposited liquid industrial waste from 209,961 tonnes to zero (MAFF, 1994). The joint international commitments were thus followed by successful implementation in all North Sea states and may accordingly be interpreted as a high degree of goal attainment internationally.

Norway has also had strong interests in strengthening regime commitments in the case of land-based emissions of nutrients and hazardous substances. In 1988 and 1989, Norway was hard hit by exceptionally toxic algae blooms. The blooms, which were linked to inputs of nutrients, killed fish in aquaculture enclosures to a value of NOK 30 million in 1988 alone. Norway urged the Paris Commission that all North Sea states should strictly follow the 50 per cent agreement on nutrients reached at the North Sea Conference in 1987 (Paris Commission, 1988). Most North Sea states have reached the 50 per cent reduction target for phosphorous inputs, while progress with regard to the reduction of nitrogen is significantly behind schedule (Ministry of Environment, 2002b). Nevertheless, the joint international commitments on nutrients have contributed to reducing Norwegian import of nutrients from Europe.

With respect to hazardous substances, Norway has given priority to their reduction in the last two decades. The evolving joint commitments on hazardous substances have thus been largely in line with Norwegian positions in the North Sea regime. Most North Sea states have made significant progress with regard to the 50 per cent and 70 per cent reduction targets agreed upon in various North Sea Declarations. A majority of North Sea states have met the 70 per cent reduction targets for mercury, lead and cadmium (Ministry of Environment, 2002b). As a result, the deposition of heavy metals along the Norwegian coast has gradually decreased. On the other hand, a growing number of new persistent organic substances are being transported by currents and winds (PCA, 2003).

The upshot of this development is that there is generally a high congruence between Norway's foreign policy positions based on Norway's interests as a net importer and the strengthening of regime commitments that have taken place.[8] Stronger international commitments on sea dumping and incineration, nutrients and hazardous substances have contributed to changing the behaviour of foreign target groups – states and non-state actors – thereby reducing the import of regulated pollutants to Norway from Europe.

Domestically, on the other hand, Norway has faced significant challenges in its efforts to achieve the various goals on marine pollution. The gap between international and domestic goal attainment is most clearly visible in the case of nitrogen. Nitrogenous substances were only reduced by around 20 per cent between 1985 and 1995. It should be noted, however, that most North Sea states have faced problems with attaining the goal on nitrogen. After 1995, reported emissions of nutrients from all coastal areas in Norway have actually increased (PCA, 2003).[9] In the area defined as 'sensitive', however, there has been a slight decrease. Since 1993, a slight decrease of input from municipal wastewater and industry has been reported, while input from agriculture has apparently remained stable (PCA, 2003).[10] Norway has now fixed a new deadline – 2005 – to achieve the 50 per cent target for nitrogen. For phosphorous substances, the 50 per cent target has been reached.

As in the case of nutrients, Norway's achievements concerning hazardous substances are mixed. On one hand, emissions of regulated chemicals known to be most dangerous have been significantly reduced. Norway has apparently been successful in reaching the targets of the Hague Declaration on hazardous substances (36 specified hazardous substances to be reduced by 50-70 per cent between 1985 and 1995) (Ministry of Environment and Energy, Denmark, 1995b).[11] The main reason for these achievements is that industrial emissions have been cut. On the other hand, there has been a general increase in the use of chemical substances and products.[12] Moreover, emissions of some substances covered by the Hague Declaration have actually increased. Since 1995, inputs of mercury to Norwegian coastal areas have increased dramatically.[13] Moreover, chemicals in spills have increased in the same period (PCA, 2003).[14] These scattered observations indicate that Norway has not succeeded in 'continuously reducing discharges, emissions and losses of hazardous substances' in line with the 1995 Esbjerg Declaration. Moreover, the environmental authorities are not satisfied with the situation:

> Reductions in emissions of certain hazardous chemicals have resulted in improvements in the state of the environment. However, heavy metals and persistent organic pollutants are still a serious problem in a number of fjords and harbours. There are also local problems in some fresh water bodies. (PCA, 2003, p.1)

It is obviously too early to judge whether Norway will reach the final target of cessation by 2020.

In sum, the congruence between environmental policy goals and the behavioural change of domestic target groups is mixed. Norway has successfully pushed for a strong regime on marine pollution internationally, but faced difficulties in attaining international and national goals at home.

Explaining Goal Attainment: The Role of Institutions

In order to explain Norwegian goal attainment internationally and domestically, emphasis will be put on the way in which the development and implementation of North Sea policy has been organised at the domestic level, the qualities of the core regime itself and not least the way other functionally linked international regimes have affected goal attainment. These three explanatory factors can be related to two different phases in line with the ambition of this book to take the latter part of the 1980s as a point of departure (see Chapter 2): In the first phase (efforts to influence regime commitments), which ran from the mid 1980s to the beginning of the 1990s, Norway was preoccupied with influencing outsiders by negotiating international commitments within the North Sea regime; in the second phase (domestic implementation) from 1992, when Norway launched its comprehensive implementation plan on the North Sea declarations, the focus changed to influencing domestic target groups in order to achieve stated national goals.

Phase One: Influencing Outsiders

The extent to which Norway's goals were attained at the international level is high in the sense that there is close correspondence between Norway's North Sea positions and the joint international commitments that have been made. Moreover, foreign target groups have actually changed their behaviour in accordance with the joint international obligations adopted, thus reducing the environmental pressure on Norway. What are the factors that resulted in this favourable outcome for Norway? Against the backdrop of the analytical framework presented in Chapter 2, we would expect that various institutional conditions must have been met: First, that domestically Norway was able to develop a coherent and ambitious foreign policy in this issue area rooted in the expertise of one ministry, notably the Ministry of the Environment; second, that Norway took on a leadership role in its efforts to affect the outcomes of the core regime, and that the regime itself became more receptive and stronger; and finally, that the goals of other international institutions functionally linked to the North-Sea regime were consistent with the objectives of Norwegian North Sea positions and the objectives of the North Sea regime.

Domestic institutions: Foreign marine policy. The Ministry of the Environment (ME) is in charge of developing Norwegian foreign North Sea policy. The Ministry of the Environment was first established in 1972, and several surveys have shown the recruitment patterns for the ME have differed significantly from those for other ministries. While ministerial employees traditionally have a background in law many ME employees have a natural science background. Moreover, they are younger, more inclined to support the environmental movement and less 'faithful' to political signals from above (Lægreid 1983; see also Reitan 1996). Bureaucrats from the water quality section within the Ministry of the Environment headed national delegations to the Oslo and Paris Commissions and participated in the INSCs. The dominance of the ME and its subordinate agencies over other Norwegian ministries in this area is clearly illustrated by the composition of the Norwegian delegation participating in the 1987 International North Sea Conference in London, which has been reckoned as the watershed in the North Sea cooperation. At this conference, the precautionary principle was adopted and decisions to cut inputs of nutrients and hazardous substances in half as well as to ban dumping and incineration at sea were made for the first time. The Norwegian delegation was composed of ten delegates and was headed by the Minister of the Environment. Five delegates came from the ME, two from the State Pollution Control Authority, one from the Institute of Marine Research and two from the Norwegian Embassy in London (Department of the Environment, UK, 1987).

The Ministry of Foreign Affairs (MFA) became involved in Norwegian foreign environmental policy after the report *Our Common Future* was debated in the UN General assembly in 1987. The report encouraged countries to develop a foreign policy for the environment. As a consequence, the MFA published *Norway's contribution to the international work for sustainable development in 1988* (Ministry of Foreign Affairs, 1988). The goals on the North Sea formulated by the MFA were fully consistent with the goals of the ME, i.e. to strengthen efforts to reduce pollution in the North Sea by means of international cooperation and to work for a ban on dumping. The MFA also became deeply involved in the UNCED process leading up to the 1992 Rio Conference, as well as the national follow-up process. However, UNCED largely evaded marine pollution in general and North Sea pollution in particular (see below).

The joint international commitments on hazardous substances and nutrients first adopted at the 1987 INSC mainly affected agriculture, industry and municipal wastewater treatment. Accordingly, the sector ministries and agencies representing agriculture, industry and local communities were important stakeholders. These interest groups were consulted prior to the 1987 conference, but they were far from prepared for the breakthrough that took place at this conference.[15] The turnabout of the UK at this conference came as a surprise for all parties (see below). Nor were the affected sectors prepared for the economic and practical consequences of implementing the joint commitments adopted in 1987. The decision-makers operated within a 'veil of uncertainty' that facilitated the adoption of joint international commitments, but would make implementation more difficult at a later stage. In addition, the principle of sector-integration was in its infancy, not only within the Norwegian environmental bureaucracy but also internationally. Representatives from the agricultural sector did not participate at the international level until the implementation problems in this sector became acute. The Norwegian State Secretary for the Ministry of Agriculture participated for the first time in the North Sea cooperation during an Intermediate Ministerial Meeting in Copenhagen in 1993, i.e. after the targets on nutrients were fixed and the national implementation plan was developed (Ministry of Environment and Energy, Denmark, 1993).

We may conclude that the Ministry of Environment initially had high responsibility in formulating Norwegian North Sea positions. Moreover, this ministry has been the 'green' pusher. Other affected governmental branches were prepared neither for the adoption of the joint international commitments, nor the consequences of these far-reaching commitments.

Core regime: Receptive and strong. The core regime on North Sea pollution is made up by the Oslo and Paris Conventions and Commissions (replaced by the OSPAR Convention in 1992) as well as the International North Sea Conferences and declarations. Norway's influence on the joint commitments produced by this regime should be understood against the backdrop of cooperation between the Nordic states. These states were all actively involved in the cooperation as importers of pollution, and none of them had any strong interest in sea dumping. The Nordic states made an effort to coordinate their positions whenever feasible. When acting together, they represented a strong coalition with significant influence on the international commitments. In the case of dumping and incineration at sea, the Nordic states aimed to phase out such activities completely. Their first joint effort to achieve this aim was presented as a joint Nordic proposal to the Oslo

Commission in 1985 (Oslo Commission, 1985). At the North Sea conference in 1987, the Nordic position prevailed and dumping and incineration in the North Sea is today history.

Norwegian and Nordic influence is less visible in the case of land-based pollution. The Nordic positions were generally in favour of stringent commitments, but were less coordinated than in the case of dumping and incineration at sea. With regard to hazardous substances, for example, Sweden and Norway have traditionally been more in favour of stringent regulation than Denmark. In the case of nutrients, Denmark was initially more ambitious than Norway and Sweden.[16] In some cases, specific proposals have conflicted with the need to protect Norwegian cornerstone industries. In the Paris Commission, for example, Norway opposed the most ambitious proposal on mercury because of the interests of one major company: Borregaard (Sætevik, 1988). Sætevik (1988) has undertaken a detailed study of the influence of various countries in the Paris Commission prior to the 1987 North Sea Conference. Taken together, the Nordic countries have had significant influence measured in terms of proposed regulations that have subsequently been adopted by the Paris Commission. Norway's influence is categorised as 'medium' when judged exclusively on its own merits.

The source of Norwegian and Nordic influence was based in a combination of knowledge and high activity in the international negotiations. Norway and the Nordic countries possessed a high degree of marine scientific knowledge, capacity and credibility.[17] In addition to marine scientific knowledge, the influence of Norway and the Nordic countries has mainly been rooted in a high level of political activity in terms of a high number of proposed regulations in the international negotiations. The high level of activity witnessed was based on a combination of material and ideal interests. As noted, the Nordic states also argued normatively in the Oslo Commission. However, the real political and intellectual leader in the North Sea cooperation was Germany (see below). Germany introduced the precautionary principle, which is seen as a precondition for the emission reduction targets adopted at the North Sea Conferences. Moreover, Germany took the initiative to arrange the first North Sea conference in 1984. The North Sea conferences proved extremely important to speed up the international cooperation on the North Sea.

Particularly one regime quality restricted the influence of the 'pushers' before the North Sea conference process. The cooperation in the Oslo and Paris Commissions was based on unanimity in practice.[18] This decision procedure provided the 'laggards' with significant veto-power over proposals that they opposed. It also led to protracted decision-making, due to the need to negotiate compromise agreements acceptable to all parties. As a result, joint international commitments were limited to measures acceptable to the least enthusiastic parties. For example, the Oslo Convention was based on unanimity, and the UK was able to block any progress until the latter part of the 1980s. As a result, decisions made within the Oslo Commission aimed to control dumping rather than reduce dumping activities.

The International North Sea Conferences significantly increased the influence of the 'pushers' by injecting political energy in the cooperation through a completely different institutional set-up compared to the Oslo and Paris

Commissions. The decision procedures within the Commissions changed from the requirement of unanimity between all parties to consensus linked to various 'fast-track' options. Previously, all 13 states bordering the North-East Atlantic had to agree on the same international regulations. Spain and Portugal do not border the North Sea, but they could not be treated differently from the North Sea states albeit other parts of the Convention area had needs different from those of the North Sea due to geographical, hydrographic and ecological circumstances. The INSCs led to a regionalisation of the cooperation and new procedures that allowed differential obligations between the North Sea and non-North Sea states. In effect, the exclusion of Spain and Portugal left the UK alone as the prime target for the other North Sea states. Important in this respect was also a change in the level of representation from low-level officials in the Oslo and Paris Commissions to ministerial representation at the INSCs that forged trade-offs reaching beyond the authority of lower level officials. This allowed the 'pushers' to intensify political pressure on the UK, which suddenly responded by accepting the precautionary principle and emission standards on dumping and hazardous substances at the 1987 INSC (Skjærseth, 2000).[19] As noted, the turnabout of the UK came as a surprise for the Norwegian delegation. In short, changes in the core regime contribute to explaining why the 'pushers' – including Norway – succeeded in shaping regime outputs on dumping and land-based emissions of pollutants to the marine environment. Note that this was accomplished in a situation characterised by significantly different material interests between the exporters and importers of pollution.

The Oslo and Paris Conventions were also significantly strengthened by the INSC process. This increased the impact of the regime on target groups and consequently reduced the environmental pressure on the net importers of marine pollution. First, the INSC Declarations were significantly more specific than previous commitments. The declarations set forth unequivocal objectives according to a baseline and deadline. The INSC Declarations of 1987 and 1990 included percentage reductions to combat emissions of hazardous substances, nutrients and dumping and incineration at sea. Second, the level of ambition increased as the INSC Declarations aimed at a significantly higher 'amount' of behavioural change than previously. Third, verification procedures and practice improved dramatically. In contrast to the Oslo and Paris Commissions, the INSCs systematically reviewed the achievements of preceding declarations by preparing comprehensive progress reports on measures taken by each country as well as on reductions in inputs of regulated substances. This change raised the level of transparency and improved compliance. Fourth, the INSC Declarations were not legally binding, but were in many cases subsequently transformed into legal decisions by the Oslo and Paris Commissions. For example, the Oslo Commission transformed the INSC declaration on phasing out dumping at sea into a legally binding decision. This increased the authoritativeness of the joint obligations and secured compliance with the aim of phasing out dumping and incineration at sea.

In short, Norway, in close alliance with the other Nordic states, had significant influence on the joint international commitments adopted within the North Sea regime. This influence was partly stimulated by the high scientific knowledge and the political activity of the Nordic states and partly by significant changes in the receptivity and strength of the core regime initiated by Germany.

Linked institutions: An avenue for venue shopping. The EU had a potential to affect the cooperation as a party to the Paris Convention and as a participant in the Paris Commission. The positions of the EU tended to diverge from the positions of the 'pushers' within the Paris Commission, as the EU Commission stood fourth as the main 'blocker' of proposals aimed at more ambitious regulation (Sætevik, 1988). This behaviour restricted the influence of those working for stronger regulation. The behaviour of the EU Commission was closely linked to the EU institutional system itself. First, the distribution of competence between the EU Commission and its member states was unclear and defined as an internal matter (Prat, 1990). However, in those cases where binding decisions were proposed in the Paris Commission, the Council of Ministers would make the final decision. Since the EU did not open for majority decisions on environmental directives until the Single European Act entered into force in 1987, EU member states not participating in OSPAR, such as Greece, could in principle block Paris Commission proposals in the Council of Ministers. North Sea states that participated both in the EU and the Paris Commission, such as the UK, could support proposals in the Paris Commission while subsequently opposing them in the Council of Ministers. Second, the 1957 Treaty of Rome did not include any explicit reference to environmental protection. However, a generous reading of related articles made it possible for the EU to develop a policy on water and marine pollution. The most significant directive concerned with water and marine pollution was the 1976 dangerous substance directive, which covers inland, coastal and territorial waters. However, the environmental policy of the EU did not gain a firm legal basis until the Single European Act. Lack of legal foundation combined with the requirement of agreement protracted EU decision making on environmental issues. For example, the EU did not succeed in adopting a dumping directive, even though the first proposal had been submitted to the Council as early as in 1976 (Suman, 1991).

The creation of the INSC process can be seen as an example of 'venue shopping' by means of establishing an alternative institutional arena. This move was undertaken by Germany, but it increased the influence on regime commitments for all states that pushed for more ambitious regulation, including Norway. Whereas the INSCs de-coupled states not bordering the North Sea from the North-East Atlantic cooperation, the declarations also proved to solve the problems of legal competence concerning the EU. The EU Commission was a party to the North Sea Conferences and Declarations, but it did not behave as the main 'blocker' in this arena. One important reason is that the INSC Declarations are based on 'soft law' that do not need formal approval by the Council of Ministers. It thus became easier for the EU to accept the North Sea Declarations since they were not legally binding on the EU members. In addition, after the 1987 Single European act, the EU's environmental policy changed profoundly: not only did the amount of legislation increase, but it also became stricter and increased in scope. From 1978 to 1987, the EU adopted an average of 16.6 items of EU environmental legislation each year, compared to 33.3 from 1988 to 1999. Between 1999 and

2001, this number had risen to an average of 94 each year (Haigh, 2002).[20] As shown in the next section, EU Environmental Directives are presently putting significant pressure on Norway.

This section can thus be concluded by noting that most, but not all, of the assumptions described above have been supported empirically. Norwegian foreign marine policy has been largely driven by one ministry. The Ministry of the Environment has prepared Norwegian positions, and officials from this ministry have participated in the international negotiations. Norway has exercised significant influence on regime commitments – particularly concerning dumping and incineration at sea in alliance with the other Nordic states. Norway took the initiative to the Oslo Convention on dumping and the Nordic states proposed to phase out such activities completely. Norway's influence increased as a result of a change in the regime itself. The International North Sea Conference process changed the 'rules of the game' in favour of the interests advocated by the pushers. This led to a strong regime that benefited the interests of the net importers of pollution. Contrary to what we could expect, the most important functionally related institution – the EU – was not in line with the position of Norway and the objective of the North Sea regime. In the first phase of the cooperation, the EU actually hampered regime progress as a result of a lack of legal competence and the requirement of agreement on marine pollution control. This problem was, however, dealt with by the establishment of the North Sea Conferences as an alternative institutional arena as well as institutional reforms within the EU itself.

Phase Two: Influencing Domestic Target Groups

In 1992, Norway presented a comprehensive plan in the form of a White Paper on how to follow up the North Sea Declarations (Ministry of the Environment, 1992). With this plan, the political attention changed from international to domestic goal attainment. While Norway was largely able to attain its goals at the international level, attaining its goals for implementation at the domestic level proved to be more difficult. The reason is that the most important target groups – industry, agriculture and municipal wastewater – did not change their behaviour sufficiently to reach the targets on nutrients and hazardous substances. There may be several reasons for this situation located at international and domestic levels (see Chapter 2): First, the will to advocate stringent joint international commitments in the core regime does not necessarily imply high incentives for a country to comply with its own commitments. Second, linked institutions established, or accessed by Norway after the targets were fixed, may have caused implementation problems for Norway. Third, and linked to domestic institutions, insufficient sector integration, inadequate policy instruments or exclusion of those actors responsible for implementation may have caused implementation problems. The remainder of this section will discuss each of these possibilities in turn.

Core regime: Pressure on domestic groups. Norway stood forth as a pusher in the North Sea regime, but the North Sea regime still had a profound impact on the behaviour of domestic target groups. As a net importer of pollution, Norway would prefer that the exporters of pollution contributed more by reducing pollution, while Norway itself contributed less. This interpretation is supported by the significant change in level of ambition from the 1985 water plan to the 1987 North Sea Conference. In 1985, the government was quite satisfied with the domestic situation (see above). In 1987 and 1990, Norway agreed to significant cut-backs in nutrients and hazardous substances that would lead to significant abatements costs. In essence, the North Sea Declaration adopted in 1987 went well beyond what Norway had preferred unilaterally in 1985. The significant change in the level of ambition from 1985 to 1987 cannot be traced back to specific events or a breakthrough in scientific knowledge. The 'algae' invasion occurred in 1988, i.e. later than the 1987 North Sea conference, and the 1987 Quality Status Report on the North Sea did not depart significantly from the 1984 report. Both reports painted a picture of high uncertainty concerning inputs, concentrations and effects (Quality Status of the North Sea, 1984 and 1987).

Norway perceived the INSC Declarations as an expression of political will – as politically but not legally binding for the North Sea states. Nevertheless, an active effort to follow up the declarations was perceived as being in the interest of Norway, since Norwegian waters are the recipient of inputs of pollution from the North Sea countries further south, both via the ocean currents and via the atmosphere. Against this backdrop,

> ... the Government underlines the importance of fulfilling our share of the commitments in the Declaration, both as part of the work to solve local environmental problems and in order to help solve common environmental problems, and possessing the necessary credibility to act as a driving force behind the efforts to ensure a high level of ambition in the international efforts to follow up the declarations. (Ministry of the Environment, 1992, p.6)

The 'soft law' INSC Declarations spurred significantly more domestic implementation activity than the previous recommendations and decisions adopted within the framework of the legally binding Oslo and Paris Conventions. According to the OECD (1993, p.49), water pollution control in Norway is '... dominated by the requirements of the North Sea Declarations agreed by all North Sea countries'. The INSC Declarations were, as noted, more demanding, more specific, more transparent, and verification of compliance immediately got high priority. In short, the INSC Declarations in their symbiotic relationship with the Oslo and Paris Commission obligations had significant impact on Norwegian implementation and goal attainment (Skjærseth, 2000).

The pathway of influence was political rather than legal. The North Sea Declarations empowered the environmental authorities to legitimise stringent action and to put pressure on the agricultural and municipal wastewater sectors. In the municipal sector, international obligations have significantly speeded up existing national plans. In 1990, the share of the population served by wastewater treatment plants in Norway was below the OECD average (OECD, 1993, p.51). Measures in the agricultural sector have to a significant extent been motivated by the INSC commitments (PCA, 1992). With regard to hazardous substances, the chief effect of the North Sea regime has been more ambitious targets and the inclusion of additional substances in domestic plans and action programs. The situation was different with regard to dumping and incineration at sea. In this case, the arrow of influence went almost exclusively from Norway to the regime, since Norway did not – with the exception of incineration at sea – have to modify its behaviour in order to meet the international obligations.

The upshot of this discussion points in two directions with regard to goal attainment. On one hand, the international obligations on nutrients in particular became more ambitious than preferred by Norway unilaterally. On the other hand, the regime had a profound impact on the behaviour of domestic target groups. As a result of these regime consequences, significantly less would have been achieved domestically in terms of emissions reduction in the absence of the North-East Atlantic/Northe Sea regime.

Linked-institutions: *New commitments and more pressure.* In the 1980s, the *vertical link* between the North Sea regime and the EU disrupted the efforts of Norway and other 'pushers' to tighten up regulations on marine pollution. The role of the EU changed significantly, however, during the 1990s when the INSC declarations were to be implemented. The EU institutional machinery has produced a comprehensive environmental policy in depth and scope: about 300 environmental directives, regulations and decisions have been adopted to date (Skjærseth and Wettestad, 2002). In the area of water and marine pollution, this development culminated with the adoption of the Water Framework Directive in 2000, which replaced seven old directives, including the directive on dangerous substances. For Norway, EU environmental policy represents a 'new institution' introduced after the most important joint commitments were fixed in the North Sea Declaration. In 1994, Norway joined the European Economic Area (EEA) Agreement with the EU, making nearly all EU environmental regulations applicable to Norway as well.

In order to understand the link between the North Sea regime, the EU and Norway, we have to understand how these different institutions have been linked. The interactive workings of the North Sea regime and the EU have proved synergetic and affected Norway as well as other North Sea states: First, the political 'soft law' INSCs have speeded up decision making within the EU. Second, the EU has facilitated domestic implementation of the INSC Declarations through its enforcement powers. These linked institutions have thus proven mutually beneficial by fulfilling different functions, all of which are needed to manage marine pollution effectively (Skjærseth, 2003).

The EU first responded to the 1987 and 1990 North Sea Declarations on inputs of nutrients. In 1991, the EU adopted two important directives based on the INSC initiatives of the late 1980s. The Urban Waste Water Directive set specific requirements on wastewater collecting systems to be implemented by the year 2000 or 2005. The Nitrates Directive aims at supplementing the above Directive by specifically addressing nutrient emissions from the agricultural sector. These directives overlap both the INSC Declarations and OSPAR commitments. The 1995 INSC Declaration links progress on implementation directly to these directives (Ministry of Environment and Energy, Denmark, 1995, p.33).

The Urban Waste Water Directive was considered in the Norwegian 1991-92 North Sea implementation plan, i.e. before the EEA was concluded and entered into force (Ministry of the Environment, 1992). Both the urban waste water and the nitrates directives have been transformed into Norwegian legislation, and they are registered as fully implemented by ESA (EFTA, 2002). The Urban Waste Water Directive has placed significant pressure on Norwegian follow-up efforts. The original 1996 regulation to implement the directive has been turned down by ESA, and a new regulation will be adopted in 2004. In terms of behavioural change at target group level, the Nitrates Directive has had some effect on livestock effluents (Korneliussen, 2003).

Second, the EU responded to the 1995 Esbjerg Declaration on the phasing out of hazardous substances. In 2000, the EU adopted the Water Framework Directive (WFD) which sets out its ambition to eliminate priority substances. Norway did not participate in the development of this directive, but the government has been actively involved in determining its practical consequences.[21] The WFD, which aims at good water quality by 2015, will have a significant impact on Norway (PCA, 2003). First, Norway has to establish new administrative borders – independent of national, regional and local administrative entities – in order to coordinate water management within new river basin districts. The ecosystem approach has already been integrated in Norway's strategy to achieve a clean sea (Ministry of the Environment, 2002a). Second, environmental targets have to be related to all water resources, including surface water, ground water and coastal water. The WFD has been coordinated with the North Sea regime in the sense that priority hazardous substances are based on those substances already included in the OSPAR convention and North Sea Declarations.[22]

The linkages between the EU and the North Sea regime have also had consequences for Norway's ability to influence regime commitments in the second phase. Norway's ability to influence EU environmental policy through the EEA agreement is limited at the outset. When Sweden and Finland allied with Denmark as EU members in 1995, the Nordic alliance in the North Sea cooperation crumbled away.[23] The accession of Finland and Sweden thus further weakened Norway's ability to influence joint international commitments on the marine environment.

Other *horizontally linked* regimes have emerged in the second phase and subsequently affected Norway's efforts to implement targets linked to international commitments on marine pollution. In the case of hazardous substances, two of the 36 substances covered by articles 2 and 3 of the 1990 Hague Declaration are also covered by the *Montreal Protocol* on the ozone layer. For these two substances (carbon tetrachloride and trichloroethane), the national goal was to phase out these substances by 1995 (see Chapter 4). Other recent international agreements that are vital to Norway's goals on hazardous substances are the 1998 Prior Informed Consent (PIC) Convention, the ECE protocols and the global POPs Convention, which was adopted in 2001. These agreements are likely to become important to Norway, since they attack transboundary pollution of hazardous substances. The North Sea Declarations cover direct discharges of nutrients as well as riverine inputs. However, indirect discharges from the atmosphere are also important for nitrogen levels in the marine environment. Accordingly, Norway's participation in CLRTAP on NO_x contributes to the aim of reducing eutrophication in the North Sea (see Chapter 5).

Other international processes evolving after the INSCs in 1987 and 1990 have mainly focused on relatively new (global) problems. Norway became actively involved in the process leading up to the World Summit in Rio through the ECE Conference 'Action for a Common Future' in Bergen in 1990. Even though marine pollution was an issue at the United Nations Conference on Environment and Development (UNCED) in Rio in 1992, the question of designing a (new) global marine convention was never really put on the agenda since a comprehensive international regulatory system was already in operation. Norway has focused accordingly – mainly on problems related to air (climate) and soil (biodiveristy). The Convention on Biodiversity (CBD) has, however, had direct consequences for the North Sea regime and Norway. In 2001, Norway ratified Annex V to the OSPAR Convention on the protection and conservation of the ecosystem and biological diversity of the maritime area. In addition to the EU Water Framework Directive, the CBD has also contributed to inspire Norway's ecosystem approach to the marine environment (Ministry of Environment, 2002a).

It may be concluded that linked regimes have been more of a help than a hinder to Norway's goal attainment domestically. The main reasons for Norway's implementation problems can hardly be found in the possibility that other regimes have directly disrupted the Norwegian follow-up processes. On the contrary, Norway would probably have been even less likely to have attained its goals domestically in the absence of other institutions that were functionally linked to the North Sea regime.

Domestic institutions: Sector integration, policy instruments and involvement. To attain national goals and fulfil the joint commitments on hazardous substances and nutrients, behavioural changes need to be made in target groups within three principal Norwegian sectors: industry, agriculture and the municipal wastewater sector. In Chapter 2, it was assumed that Norway's ability to change the behaviour of domestic target groups in accordance with stated goals would depend on the degree of sector integration (horizontal and vertical), the adequacy of policy instruments, and access to decision-making at an early stage for those actors responsible for implementation.

Norway has made significant progress on regulating hazardous substances, but major challenges still remain. Direct discharges from industry have traditionally been the main source of input to the marine environment of hazardous substances. Such point sources are dealt with by a simple administrative system: A single authority, the Norwegian Pollution Control Authority (PCA), and one principal pollution law, the 1981 Pollution Control Act that covers land, air and water pollution. The Ministry of the Environment has primary responsibility for this law. The risk of disintegration within the state apparatus in this sector is thus limited.[24]

The principal policy instrument applied to industrial point sources is *legislation* in the form of discharge permits based on the Pollution Control Act.

The PCA monitors compliance in the industrial sector through reports from companies, inspections and source testing. The permit system is generally given credit for the achievements witnessed (Ministry of the Environment, 1995). The permit system has since 1981 been based on Integrated Pollution Control (IPC), which has been perceived as a major advantage by industry compared to other countries where separate control authorities exist for water, air and land. Systematic verification of compliance and the use of sanctions in clear cases of violation have contributed to the results achieved. For example, from 1989 to 1990 the PCA fined lack of compliance in 52 cases; in 48 of these, the companies took action within the time limits set (Ministry of Environment, 1995). In 1995, 395 firms were controlled and a total of 69 violations and irregularities were detected, of which 43 were followed up administratively, nine were prosecuted, and 17 were made subject to fines (PCA, 1996). In essence, the progress made on hazardous substances could be achieved by including new substances and obligations in an existing permit system. It is quite illustrative that the sector target for industry has almost been reached in the case of nitrogen emissions, in contrast to the municipal and agricultural sectors (OAGN, 2000).[25]

Since 1995, there are indications that the control system has not kept pace with the increase in permits. According to the Office of the Audit General of Norway (OAGN), significant shortcomings have been detected in the control system for the industrial sector (OAGN, 2000). These shortcomings have to be dealt with in order to reach the target on continuously reducing discharges, emissions and losses of hazardous substances. From 1995 to 1998, the number of land-based companies with emission permits almost doubled.[26] About 40 per cent of those companies controlled by the authorities in 1997 and 1998 had not complied with their permits. This indicates a significant increase in violations compared to 1995. The OAGN hold that in cases of non-compliance, the benign enforcement that has been used has not proved sufficient to reduce the number of incidents of non-compliance. The OAGN recommends that stronger enforcement, including fines and prosecution, should be considered.

Norway has not even come close to fulfilling the targets on nitrogenous substances. The goals on nutrients demand substantial efforts within the agricultural and municipal waste-water sectors, in addition to industry. The relationship between the Ministry of the Environment and the Ministry of Agriculture is crucial for Norwegian goal attainment on nutrients. The 'sector responsibility' approach was introduced in the 1988-89 White Paper to Parliament on the follow-up to the World Commission on Environment and Development (Ministry of the Environment, 1989). The aim was to integrate environmental concerns into other sectors, such as agriculture. Sector responsibility applied in Norway gave all relevant ministries responsibility for carrying out a sustainable policy within their respective sectors. Several steps were taken, including integrated planning and the establishment of inter-ministerial committees. The principle of 'sector responsibility' gives the Ministry of Agriculture and subsidiary departments the main responsibility for agricultural pollution matters. However, the main legislation concerning agricultural pollution is the Control of Pollution Act from 1981 which is administered by the environmental authorities.

The Ministry of Agriculture started to pay serious attention to environmental issues in the mid 1980s. Since then, environmental awareness has developed gradually. Since 1987, meetings on agricultural policy have been held twice a year by the agricultural and environmental ministries. In recent years, this contact has been further formalised and meetings have become more frequent. In 1989, the ME was for the first time included in the national committee responsible for agricultural support (e.g. subsidies). Similar development has taken place at the regional level, and coordination between environmental departments and agricultural offices at this level has increased significantly (Mydske and Steen, 1994). According to the Ministry of Agriculture, the North Sea Declarations place strong obligations on agriculture, and the Ministry places significant emphasis on following up the international commitments (Ministry of Agriculture, 1992).

Whereas the degree of sector integration has increased, it has apparently been too little too late. There are also indications of recession since 1995. For example, new data on inputs of nutrients have not been produced by the Ministry of Agriculture since 1996 (OAGN, 2000). This makes it difficult to identify and evaluate the effectiveness of policy instruments. Changing the behaviour of farmers has proved difficult, and implementation problems in this sector stem partly from inadequate policy instruments. Policy instruments have mainly been based on *general legislation* backed up by weak verification and enforcement. Due to the high number of targets, the effectiveness of general legislation rests heavily on the 'good faith' of farmers.

A tax on commercial fertilizer was introduced in Norway in 1988. In the Norwegian implementation plan on the North Sea Declaration, it was recommended that this tax be increased substantially in order to have an effect on consumption of fertilizer in the short term (Ministry of the Environment, 1992). Taxes have been opposed by the Norwegian Farmer's Union, and the tax on artificial fertilizer was not increased; it was actually abandoned in 1999. Some measures have also been implemented in 'wrong' geographic areas. New cultivation practices – that were identified as one major measure in the implementation plan – have mostly been applied outside the area defined as 'sensitive' where the 50 per cent reduction target applies (OAGN, 2000). However, there has also been some progress on policy instruments and measures in the agricultural sector. As a consequence of the INSC Declarations, new measures on diffuse run-offs and soil erosion have been introduced.

Municipalities have traditionally been responsible for several issue areas with relevance for environmental policy. Of particular importance here is a law from 1974 that permits municipalities to finance all investments in wastewater treatment through local taxes, leaving them significant discretion. When state initiatives are based on the Control of Pollution Act, the state may overrule the priorities of the municipalities. On the other hand, the state is cautious about using this authority in practice, due to the principle of local autonomy. Local conditions and competence were taken seriously into account when the Pollution Control Authority developed its plan on how to achieve the 50 per cent goals on nutrients, particularly concerning the municipal sector (PCA, 1992). Many municipalities had lagged far behind the needs as defined by the central government, which had specified that activity should be speeded up considerably (Ministry of the Environment, 1992). As noted, the INSC Declarations thus became an instrument for the ME to put pressure on the municipalities.

Municipalities are responsible for necessary purification according to the Pollution Control Act and the principle of 'sector responsibility'. The state, in the capacity of the County Governor, lays down specific requirements concerning drainage and purification in the permits granted. However, the high expectations in Oslo have been dashed at municipal level. All affected municipalities submitted appeals to the Ministry of the Environment concerning the obligations for nitrogen removal issued by the county governor offices. Eventually, fierce opposition at local level contributed significantly to the cancelling of planned investments at 24 plants (Skjærseth, 1999). Moreover, municipal wastewater treatment plants operate with emission permits issued by the County Governor. In 1997 and 1998, 20 municipal treatment plants were controlled, and non-compliance with the emission permits was detected in 85 per cent and 55 per cent of the cases respectively (OAGN, 2000).

In addition to insufficient sector integration and inadequate policy instruments, the implementation problems witnessed in the municipal and agricultural sectors are also directly related to the link between foreign and national North Sea policy. We have to go back to the original 1987 goal of reducing nutrients to *sensitive areas* in order to fully understand Norway's implementation problems concerning nitrogen. National authorities had responsibility for defining the sensitive areas to which the 50 per cent reduction target would apply. Since Norway had urged other North Sea states to strictly follow this goal in the wake of the algae 'invasion', Norway felt obliged to define a relatively large coastal area in the south-eastern part of the country as sensitive (in contrast to, e.g., the UK, which subsequently argued that there were no 'sensitive' areas in the UK). Norway thus became 'trapped' in its own international ambitions. The wide definition of sensitive areas significantly increased Norway's abatement costs.

As noted, the Norwegian goals on nutrients also required new policy instruments and involvement by the Ministry and agencies of agriculture, farmers' organisations as well as local communities represented by the Ministry of Local Government and Regional Development. These interests initially did not see the economic and practical consequences of Norway's international commitments on nutrients. Even though agricultural and wastewater interests were consulted, the ME had its 'core insiders' that were regularly informed and invited to discuss various matters related to the North Sea conferences. The most important environmental organisations and some target group organisations were regularly informed and consulted, including business interests. The farmers' organisations and representatives for municipalities did not belong to the 'core insiders' (Skjærseth, 1999).[27] The positions on North Sea policy among these organisations have varied accordingly: Norwegian industry organisations have explicitly supported implementation of the INSC Declarations, municipalities have opposed them, and the Norwegian Farmers' Union has been reluctant. A higher level of goal attainment could perhaps have been attained if these interests had been more actively included earlier in the decision-making process.

Domestic implementation problems represent an important explanation of the relatively low level of domestic goal attainment witnessed. These problems have mainly been caused by insufficient sector integration and inadequate policy instruments. In addition, some target groups, agencies and ministries responsible for implementation were barely included in the decision-making process and were initially not aware of the economic and practical consequences of the targets on nitrogen in particular.

Alternative Explanation: Problem Types

Institutions at domestic and international levels are not all that matter. This section explores the merits of a complementary explanatory approach that takes problem types as a point of departure. As noted in Chapter 2, problems are most 'malign' and the prospects for goal attainment most bleak in situations characterised by opposing and influential member-states in the core regime and the presence of opposing target groups at home. Strong opposition can be expected if the costs are concentrated to specific target groups while the benefits are widely distributed throughout society. Conversely, problems are most 'benign' and the prospects for goal attainment most bright in situations characterised by compatible interests with influential states abroad and supporting target groups at home. Strong support can be expected if the benefits are concentrated to specific target groups whereas the costs are widely distributed.

Let us commence with the interests and positions of other member states in the core regime. The most powerful North Sea state – in terms of marine pollution related capabilities – has actually been the UK, which has generally opposed Norwegian positions. The UK had significantly higher marine scientific capacity than all the Nordic countries combined (Skjærseth, 1991). As noted, the break-through at the 1987 North Sea Conference can best be understood in light of leadership and institutional reforms, which opened for political pressure directed towards the UK. In this process, the Nordic countries played an important role. This is most clearly visible in the case of dumping at sea, where the EU did not have any legislation and the UK stood fourth as the main 'dumper'. Moreover, most continental North Sea states were involved in this practice. In this case, the positions of Norway and the Nordic countries prevailed even though they faced a quite 'malign' problem in political terms.

The situation for land-based emissions is more complex: the Nordic positions were more mixed according to specific regulations and substances. Denmark was the main force behind the ambitious commitments on nutrients, while Norway and Sweden pushed for stringent action on hazardous substances. As for dumping, the UK was the major opponent to the Nordic states. However, the important role played by Germany proved instrumental for Norway and the Nordic states. Germany was the intellectual leader by introducing the precautionary principle and the entrepreneurial leader by initiating the North Sea conference process. These actions formed the knowledge base and the institutional framework that came to serve as a necessary condition for Norway's high level of goal attainment internationally. The upshot of this brief discussion is that Norway's high level of goal attainment occurred both in alliance with (land-based), and in opposition to other influential North Sea states (dumping). Overall, the leadership role played by Germany and Norway's alliance with other Nordic states stacked the deck in Norway's favour.

Domestically, the implementation problems faced by Norway can partly be explained by opposing target groups at home arising from high abatement costs. As noted, joint decisions were initially made under a 'veil of uncertainty', and high abatement costs for agriculture and municipalities surfaced mainly after the joint international commitments were adopted. In 1992, total investment costs for implementing the North Sea Declarations were estimated at NOK 10 billion between 1985 and 1995 (Ministry of the Environment, 1992). The size of these costs was, however, to a large extent determined by political processes at the interface between the international and domestic levels that resulted in a wide definition of 'sensitive areas' (see above).

If we look at how these costs were allocated at the target group level, we see that the cost/benefit way of reasoning does not sufficiently explain what happened and why. Industry faced concentrated costs and widespread benefits to the society at large and had thus incentives to oppose the North Sea Declarations. However, industry actually supported Norwegian goals and delivered substantial behavioural change on hazardous substances and even on nitrogenous substances. Agriculture had in principle the same incentives as industry, but reluctantly supported the targets and delivered low-level behavioural change on nitrogen. Norwegian farmers are heavily subsidized at the outset, and a significant share of the abatement costs are actually paid by the state. In this perspective, we would have expected that agriculture would have been more supportive than industry from a purely cost/benefit line of reasoning. The municipal wastewater sector deviates from industry and agriculture in the sense that abatement costs are paid through local water charges. In addition, state subsidies to municipalities have been an important economic tool. Nevertheless, all affected municipalities strongly opposed obligations for nitrogen removal issued by the county governor offices by submitting appeals to the Ministry of the Environment.

The difficult problem type in terms of resistance from target groups can contribute to our understanding of the relatively low level of domestic goal attainment, particularly with regard to nitrogen. The reasons for this opposition can, however, only to a limited extent be traced back to the distribution of costs and benefits. Moreover, the size of the abatement costs was determined by a political process in which Norway became 'trapped' in its international ambitions. In short, the difficult problem type does not carry sufficient explanatory power to explain the outcome. Institutional factors such as regime pressure, sector-integration, and the adequacy of policy instruments appear equally, if not more, important to understanding the achievements as well as the shortcomings witnessed.

Conclusion

Norway's efforts to strengthen the international regime and thereby affect the behaviour of other North Sea states have proved successful in terms of goal attainment. The clearest example is the Oslo Convention on dumping and incineration at sea. Norway took the initiative to establish this convention, which has contributed to a phase out of dumping activities in the North Sea and the wider North-East Atlantic area. Norway's goal attainment domestically has, however, been mixed. On one hand, Norway has made significant progress on water and marine pollution control. Regulated hazardous substances have been reduced significantly, and emissions of phosphorous substances have been cut by 50 per cent in sensitive areas against the 1985 baseline. On the other hand, Norway has not even come close to reaching its goal on nitrogenous substances, and there are indications that the achievements witnessed on hazardous substances have stagnated or have been going in the wrong direction according to stated goals. This pattern implies that Norway has achieved a higher level of goal attainment abroad than at home.

The main reasons for Norway's high goal attainment at the international level are linked to cooperation with other Nordic states and changes in the receptivity and strength of the core regime, as well as in the linkage between the North Sea regime and the EU. Norway's North Sea foreign policy has been in the hands of the Ministry of the Environment, which has generally pushed for stringent joint commitments in alliance with other Nordic states whenever feasible. However, initial Nordic efforts to strengthen regime commitments were blocked in the first phase mainly by the UK and the EU Commission due to the requirement of unanimity in practice.

Germany's introduction of the precautionary principle and its initiative to establish the International North Sea Conferences led to a change in regime receptivity that benefited the interests of Norway and the other Nordic states. The decision rules changed from unanimity to consensus linked to various 'fast track' options. Particularly important in this respect was that joint commitments could be adopted by the North Sea states only. Other parties bordering the wider North-East Atlantic, such as Spain and Portugal, could as a result be treated differently than the North Sea states. This change opened up for political pressure directed towards the UK, which responded by modifying its position. As a consequence, the regime became significantly stronger in terms of level of ambition, specificity, verification procedures and transparency.

Change in the receptivity of the North Sea regime was also closely linked to change in the linkage with the EU. The EU Commission as a party in the cooperation had slowed down progress due to a lenient environmental policy that did not have a clear legal foundation in the Treaty of Rome. In this situation, the establishment of the INSC process serves as an example of 'venue shopping' by establishing a new and partly overlapping institution to the Oslo and Paris Conventions and Commissions (later replaced by the OSPAR Convention). The INSC Declarations were not legally binding and could be supported by the EU without formal decisions in the Council of Ministers. In this way, the INSC

Declarations bypassed the legal obstacles linked to the EU. Moreover, from the late 1980s, the EU as an institution changed significantly in legal basis and decision procedures and the EU rapidly picked up momentum in environmental policy that subsequently came to place strong pressure on Norway's implementation of marine pollution policies.

One major reason why Norway in the 1990s has made progress on domestic goal attainment lies in the pressure from the core regime as well as linked institutions. Norway would probably have achieved significantly less domestically in the absence of this institutional pressure. The North Sea regime itself had a significant impact on Norwegian marine policy and implementation. The main pathway of influence appears to be that the environmental authorities used the international North Sea declarations to put pressure on reluctant domestic sectors, i.e. agriculture and municipal wastewater that had traditionally lagged behind the targets set. This change in the direction of influence – from Norway's efforts to strengthen regime commitments to regime impact on Norway – needs to be understood in a dynamic time perspective. Norway as a net importer of pollution would prefer that others, i.e. the exporters, contributed more to emission reduction, while Norway itself contributed less. Norway increasingly saw itself as a 'victim' of transnational pollution and the implementation and compliance of others were crucial to Norway's goal attainment and environmental quality.

The EU is another reason behind the pressure from international institutions brought to bear on Norway. In 1994, Norway entered the EEA agreement, which made most EU directives and regulations on water and marine pollution binding for Norway. As noted, EU environmental policy in general and marine and water policy in particular developed significantly through the 1990s in scope, stringency and depth. EU directives are presently seen as an important tool to implement the North Sea declarations, particularly in the agricultural and wastewater sectors. The EU Water Framework Directive will have a significant impact on Norwegian water and marine policy in the future, with its emphasis on ecosystems as the basic unit for regulation.

In spite of significant pressure from international institutions, Norway has experienced implementation problems at home. These problems can first be linked to insufficient sector integration horizontally, particularly in the agricultural sector. Even though the agricultural authorities have increasingly integrated water and marine pollution concerns, it has been too little too late. Secondly, there has been insufficient integration of environmental concerns vertically, i.e. between central and local administrative levels. Norwegian municipalities have significant competence in wastewater treatment, but this competence has been used to oppose requirements of nitrogen removal facilities. Thirdly, the policy instruments applied have sometimes been inadequate. The permit system applied on industry and municipalities has apparently become less effective since the mid 1990s, and stronger enforcement has been suggested in order to increase compliance with the permits issued. In the agricultural sector, policy instruments have been applied in 'wrong' geographical areas, and the tax on commercial fertilizer was abandoned rather than increased. Finally, those authorities and actors responsible for implementation were not aware of the consequences and were barely involved in the decision-making process when the targets were fixed internationally. This has probably contributed to reluctant behaviour in the implementation phase both in the agricultural and wastewater sectors.

During the 1990s it became more difficult for Norway to influence international commitments on the marine environment, due to the increasing role of the EU and the withering of Nordic cooperation in the wake of Sweden and Finland's accession to the EU. However, in line with the main principles of the Water Framework Directive, Norway has recently developed a new domestic strategy on the management of the marine environment. The ecosystem approach aims to take a more holistic perspective on regulation in which the ecosystem itself serves as the prime unit across administrative borders. If implemented, this reform may improve the disintegration witnessed between various affected sectors. This will be needed for Norway to deal with the significant challenges of meeting the aim of zero emissions of the most hazardous substances by 2020.

The observed gap between international and domestic goal attainment formed the point of departure for this study. This gap would have been wider and the emissions of polluting substances significantly higher in the absence of the impact of international institutions on Norway.

Notes

1 Land-based (river input and direct discharge) and ocean-based discharges (dumping and incineration at sea) of hazardous substances and nutrients (phosphorous and nitrogenous substances) as well as atmospheric fall-out have been among the major sources of contaminants to the North Sea. Inputs of hazardous substances can cause a number of problems to the environment and human health since many hazardous substances are persistent and break down very slowly in the environment. Emissions of nutrients cause algae blooms and oxygen depletion in the marine environment.

2 However, Norway did deliver waste for incineration at sea up to 1989.

3 Two executive commissions – which met annually – were set up in order to implement and review the functioning of the Conventions. The Oslo Commission (OSCOM) and the Paris Commission (PARCOM) were assisted by the Standing Advisory Committee for Scientific Advice (SACSA) and the Technical Working Group (TWG) respectively, by ad hoc working groups and by the Joint Monitoring Group (JMG).

4 From 1990, Switzerland was also invited to participate.

5 The 1987 Single European Act incorporated environmental protection into EU legislation, the 1991 Maastricht Treaty and the 1997 Amsterdam Treaty changed EU environmental decision-making towards a wider application of qualified majority.

6 Defined as the coastline between the Swedish border and Lindesnes.

7 See Ministry of the Environment (1992) and (2002a).

8 'Generally' in the sense that there are also examples showing that Norway has not been in the forefront for stronger regulation on specific issues. One example is dumping of abandoned offshore installations in the North Sea.

9 From 101,323 tonnes in 1995 to 118,174 tonnes in 1999 (Pollution Control Authority, 2003).

[10] Discharges of nutrients from aquaculture were included from 1998 (Pollution Control Authority, 2003).

[11] According to the environmental authorities, however, data on emissions of hazardous substances before 1994 are less representative.

[12] In Norway, there are about 8,000 to 10,000 chemical substances in about 50,000 products (Pollution Control Authority, 2003).

[13] The significant increase came in 1998: From 489 kg in 1997 to 2,839 kg in 1998 and 2,151 kg in 2000. According to the Pollution Control Authority (2003), the significant increase from 1997 to 1998 is hard to explain and different methods for measuring episodes of discharges have been tried out at present.

[14] From 420 Unit m^3 in 1995 to 930 Unit m^3 in 2000 (Pollution Control Authority, 2003).

[15] Personal communication with Per Schive 12 September 2003. Per Schive participated in the Norwegian delegation at the 1987 INSC.

[16] Personal communication with Per Schive 12 September 2003.

[17] This is measured in terms of the number of marine scientists, publication of marine scientific peer reviewed articles as well as the number of research programmes directed at the North Sea in 1990. See Skjærseth (1991).

[18] The Paris Convention included qualified majority, but unanimity came to serve as the basic decision rule in use.

[19] The North Sea regime led the UK to adopt the so called 'Red List' in 1989 identifying the sources of 23 specific substances.

[20] These numbers do not distinguish between important legislative initiatives and minor adjustments, improvements or amendments of existing legislation.

[21] Personal communication with Per Schive 12 September 2003.

[22] The WFD lists 33 specific hazardous substances.

[23] Personal communication with Per Schive 12 September 2003.

[24] However, from 1993 regulation of smaller industrial plants was delegated to the county level.

[25] Industry reached 71 per cent reduction in 1999 compared to a sector target amounting to 75 per cent.

[26] From about 240 to 430 companies.

[27] The environmental organisations regularly informed and consulted were the following: the Norwegian Society for Conservation of Nature (NSCN), Bellona, Nature and Youth, and Greenpeace Norway. The target groups were the Norwegian Ship Owners' Association, the Norwegian Fisherman's Association and the Federation of Norwegian Industries (FNI) and the Norwegian Confederation of Business and Industry (NCBI from 1989).

References

Department of the Environment (1987), *Summary Record of the Second International Conference on the Protection of the North Sea, London, 24-25 November* (Summary Record), London.

EFTA (2002), AIDA – Acquis Implementation Database (accessed 8 August 2002): www.efta.int/structure/SURV/efta-srv.asp.

Ehlers, P. (1990), 'The History of the International North Sea Conferences', in D. Freestone and T. IJlstra (eds), *The North Sea: Perspectives on Regional Environmental Co-*

operation, Special Issue of the *International Journal of Estuarine and Coastal Law*, Graham & Trotman, London, pp. 3-15.

Haigh, N. (2002), *Manual of Environmental Policy: The EC and Britain*, Cartermill International, London, 1992 and subsequent revisions.

Korneliussen, Y. (2003), *Implementing EU Environmental Directives in Norway: Two Cases of Water Pollution*, University of Oslo, Department of Political Science, Oslo, Spring.

Lægreid, P. (1983), 'Miljøverndepartementet. Ny organisasjon i etablert miljø', *Nordisk Administrativt Tidsskrift*, nr. 4, pp. 365-83.

MAFF (1994), *Monitoring and Surveillance of Non-Radioactive Contaminants in the Aquatic Environment and Activities Regulating the Disposal of Wastes at Sea, 1992*, Ministry of Agriculture, Fisheries and Food, Directorate of Fisheries Research, Lowestoft.

Ministry of Agriculture (1992), *St.prp.nr. 8 1992-93*, Oslo.

Ministry of the Environment (1975), *On the National Plan for Utilization of Water Resources* (Om arbeidet med en landsplan for bruken av vannressursene), Report No. 107 to the Storting (1974-75), Ministry of the Environment, Oslo.

Ministry of the Environment (1985), *On Measures Against Water- and Air Pollution and Municipal Waste (Om tiltak mot vann- og luftforuresninger og om kommunalt avfall)*, Report No. 51 to the Storting (1984-85), Ministry of the Environment, Oslo.

Ministry of the Environment (1989), *Environment and Development. Programme for Norway's Follow-up of the Report of the World Commission on Environment and Development (Miljø og Utvikling. Norges oppfølging av Verdenskommisjonens rapport)*, Report No. 46 to the Storting (1988-89), Ministry of the Environment, Oslo.

Ministry of the Environment (1992), *Concerning Norway's Implementation of the North Sea Declarations*, Report No. 64 to the Storting (1991-92), Ministry of the Environment, Oslo.

Ministry of the Environment (1995), *On Instruments in Environmental Policy*, NOU report, Ministry of the Environment, Oslo.

Ministry of the Environment (2002a), *Clean and Abundant Sea (Rent og rikt hav)*, Report No. 12 to the Storting (2001-02), Ministry of the Environment, Oslo.

Ministry of the Environment (2002b), *Ministerial Declaration of the Fifth International Conference on the Protection of the North Sea, Bergen, Norway 20-21 March 2002* (Bergen Declaration), Ministry of the Environment, Oslo.

Ministry of Environment and Energy (1993), *Statement of Conclusions from the Intermediate Ministerial Meeting 7-8 December 1993 in Copenhagen* (Statement of Conclusions), Danish Environmental Protection Agency, Copenhagen.

Ministry of Environment and Energy (1995a), *Fourth Ministerial Conference on the Protection of the North Sea, Esbjerg, Denmark, 8-9 June* (Esbjerg Declaration), Danish Environmental Protection Agency, Copenhagen.

Ministry of Environment and Energy (1995b), *Fourth Ministerial Conference on the Protection of the North Sea, Esbjerg, Denmark, 8-9 June, Progress Report*, Danish Environmental Protection Agency, Copenhagen.

Ministry of Foreign Affairs (1988), *Environment and Development: Norway's Contribution to International Efforts for Sustainable Development* (Miljø og utvikling: Norges Bidrag til det internasjonale arbeid for en bærekraftig utvikling), Ministry of Foreign Affairs, Oslo.

Mydske, P.K. and Steen, A., in association with Taarud, A. (1994), 'Land-use and Environmental Policy in Norway', in K. Eckberg, P.K. Mydske, A. Niemi-Iiahti and

K.H. Pedersen (eds), *Comparing Nordic and Baltic Countries—Environmental Problems and Policies in Agriculture and Forestry*, Nordic Council of Ministers, Copenhagen, TemaNord, p. 572.

OAGN – The Office of the Auditor General of Norway (2000), 'Riksrevisjonens undersøkelse av Norges oppfølging av OSPAR-konvensjonen innen industri-, avløps- og landbrukssektoren', Document nr. 3, p. 4 (2000-2001), Oslo.

OECD (1993), *Environmental Performance Reviews: Norway*, OECD Publications, Paris.

Oslo Commission (1985), *Tenth Annual Report of the Oslo Commission*, Secretariat for the Oslo and Paris Commissions, London.

Paris Commission (1988), *Summary Record of the Tenth Meeting of the Paris Commission, Lisbon, 15-17 June*, Paris Commission, London.

PCA, Norwegian Pollution Control Authority (Statens forurensningstilsyn, SFT) (1992), *Ministerial Declaration on the Protection of the North Sea. Analysis of Measures to Reduce Nutrient Inputs*, Report 92, p. 34, Oslo.

PCA, Norwegian Pollution Control Authority (Statens forurensningstilsyn, SFT) (1996), *SFTs kontroll-resultater 1995*, Report 96, p. 12, Oslo.

PCA, Norwegian Pollution Control Authority (Statens forurensningstilsyn, SFT) (2003), 'State of the Environment Norway', www.environment.no and www.sft.no/arbeidsomr/vann/vanndirektiv/dbafile6858.html.

Prat, J.L. (1990), 'The Role and Activities of the European Communities in the Protection and the Preservation of the Marine Evironment of the North Sea', in D. Freestone and T. IJlstra (eds), *The North Sea: Perspectives on Regional Environmental Co-operation*, Special Issue of the *International Journal of Estuarine and Coastal Law*, Graham & Trotman, London.

Quality Status of the North Sea 1984 (1986), A report compiled from contributions by experts of the governments of the North Sea Coastal States and the Commission of the European Communities prepared for the International Conference on the Protection of the North Sea, Bremen, October 31 to November 1, 1984, *Deutsche Hydrographische Zeitschrift*, no. 16, Hamburg.

Quality Status of the North Sea (1987), Summary, Second International Conference on the Protection of the North Sea, Scientific and Technical Working Group, HMSO, Department of the Environment, London.

Reitan, M. (1996), *Norway: A Case of 'Splendid Isolation'*, Department of Political Science, University of Oslo.

Skjærseth, J.B. (1991), *Effektivitet, problem-typer og løsningskapasitet: En studie av Oslo-samarbeidets takling av dumping i Nordsjøen og Nordøstatlanteren*, R:009, The Fridtjof Nansen Institute, Lysaker.

Skjærseth, J.B. (1999), *The Making and Implementation of North Sea Pollution Commitments: Institutions, Rationality and Norms*, Department of Political Science, University of Oslo.

Skjærseth, J.B. (2000), *North Sea Cooperation: Linking International and Domestic Pollution Control*, Manchester University Press, Manchester.

Skjærseth, J.B. (2003), 'Protecting the North-East Atlantic: Enhancing Synergies by Institutional Design', paper presented at the 44th Annual ISA Convention, Portland, Oregon, February 26 to March 1, 2003.

Skjærseth, J.B. and Wettestad, J. (2002), 'Understanding the Effectiveness of EU Environmental Policy: How Can Regime Analysis Contribute?' *Environmental Politics*, vol. 11, no. 3, pp. 99-120.

Suman, D. (1991), 'Regulation of Ocean Dumping by the European Economic Community', *Ecology Law Quarterly*, vol. 18, no. 3, pp. 559-618.
Sætevik, S. (1988), *Environmental Cooperation between the North Sea States*, Belhaven Press, London.

Chapter 7

Climate Change: Cost-effectiveness Abroad, Possibilities at Home[1]

Hans-Einar Lundli and Marit Reitan

Introduction

The policy of establishing an international climate regime can be traced back to the second half of the 1980s. An international climate treaty entered into force in March 1994, and a binding agreement to reduce greenhouse gas emissions was agreed upon in Kyoto in December 1997. However, for the Kyoto Protocol to enter into force, Russia has to ratify the treaty, and, at the time of this writing, the Russian authorities have not made it clear whether they will ratify or not. A treaty with binding emission commitments is therefore not yet in force, and the target time period for emissions reductions is still in the future (2008-12).

At the national level, Norway was a pioneer in the early 1990s in developing a climate change policy. It became the first country to adopt a national emission target in 1989, and was also a forerunner in adopting a tax on emissions of carbon dioxide (CO_2) in 1991. However, the main purpose of being a forerunner in this field was to trigger the process of establishing an international climate treaty with *binding* emission commitments. In 1997, the unilateral stabilisation target was replaced with the national commitment agreed upon in Kyoto that allowed Norway to increase its emissions of six greenhouse gases by one per cent (measured in CO_2 equivalents) from 1990 to 2008-12.[2]

The aim of this chapter is to explain Norway's goal attainment with respect to climate policy objectives at both the national and international levels. We argue that there has been a high degree of overlap between the Norwegian negotiation positions and the establishment of a climate regime with differentiated emission targets between countries and possibilities for emission trading and joint implementation.[3] We highlight three factors when explaining this: The emphasis on cost effectiveness as an overriding principle in environmental politics; increased knowledge about the flexibility mechanisms; and the decision rules of the climate regime combined with the high number of participants. With respect to the national dimension, we look particularly at how the development of the flexibility mechanisms reduced conflicts between political parties and government ministries by promising low implementation costs. We then investigate to what degree a lack of coordination between the external and internal dimension of Norwegian climate policies can have affected the Norwegian goal attainment. Finally we discuss the malignancy of the climate problem and how this affects Norwegian goal attainment.

Norway's Goal Attainment in the Field of Climate Change

Revised Domestic Goals Improved Outlook for Successful Implementation

When the Norwegian Parliament discussed the national strategy to follow up the recommendations from the Brundtland Commission in spring 1989, the Brundtland government proposed a goal of stabilising CO_2 emissions at the national level by the year 2000.[4] The Prime Minister was strongly criticised for proposing a target that according to the recommendations given by the Brundtland Commission was far from sufficient. Environmental issues were at the top of the political agenda, and the parliamentary debate escalated, resulting in a goal of stabilization by the year 2000, using 1989 as the base year. The debate was carried out within an atmosphere described as a 'green beauty contest' to propose the most ambitious climate measures (Reitan, 1998). However, the stabilization target included some important reservations that often are neglected. From the outset it was stated that the stabilization target was a *preliminary* target that had to be evaluated continuously in relation to *technological development* and to future international climate change *negotiations and agreements*. National climate efforts were, according to the government, meaningful only to the extent that they helped accelerate an international process in this field (Reitan, 1998, p.121).

By the mid 1990s the Parliament had to announce that the goal was far too ambitious (Reitan, 1998, pp.146-7). Not only was there no international climate treaty with binding emission reductions in place; Norwegian authorities were unwilling to adopt ambitious unilateral measures that would impose considerable costs to industry and trade. The stabilization target remained the overall national goal until the end of 1997, but Norway was far from reaching it. In 2000, CO_2 emissions were 20 per cent higher than 11 years earlier (PCA, 2000). This development was, however, no surprise. In the White Paper following up the Brundtland Report, estimates projected a 28 per cent increase in emissions from 1989 to 2000 if no measures were adopted (Ministry of the Environment, 1989, p. 58).

In 1997 the unilateral stabilization goal was replaced by the national commitments agreed upon in Kyoto. This not only implied a less stringent target, but also a possibility for Norway to meet its targets by choosing from several different greenhouse gases to reduce. While the national stabilization target covered only CO_2, the scientific knowledge of anthropogenic climate change had progressed during the first half of the 1990s and the importance of reducing also other types of greenhouse gases became more apparent. An index called the Global Warming Potential Index (GWP) was developed, making it possible to compare the effect on the climate of reducing different types of greenhouse gases (PCA, 1997, p.8). The Kyoto target thus covers CO_2 and five additional greenhouse gases (or groups of gases): methane (CH_4), nitrous oxide (N_2O), perfluorocarbons (PFCs), sulphur hexafluoride (SF_6) and hydroflourocarbons (HFCs).

At the outset, the prospects for Norway of meeting its Kyoto target did not seem to be good. A business-as-usual scenario indicates that the emissions of greenhouse gases could increase by 24 per cent from 1990 to 2010 (PCA, 2000). If the planned construction of three gas-fired power plants is realized before 2010, domestic emissions of greenhouse gases could increase by nearly 30 per cent. The government's strategy to achieve emission reductions is to develop an emission permit system that will become the backbone of the Norwegian climate policy in the years ahead (Ministry of the Environment, 2002). It is to be fully implemented from 2008, replacing carbon taxes as the main policy instrument in Norway's climate policy. Within a national emission permit system, a cap on the total emissions of greenhouse gases in Norway is set (corresponding to the Kyoto target). The quotas (or emission certificates) allocated to Norway in accordance with the Protocol are then are then sold or distributed as permits to the entities/companies to be included in the permit system. A permit gives the holder the right to emit a certain amount of greenhouse gases within a limited time span, and permits can be bought and sold between the different entities. The national permit system is to be fully connected to the flexibility mechanisms of the Kyoto Protocol. This means that the entities – either private companies or government agencies – can buy quotas from abroad to meet their emission target.

The present centre-right Government emphasizes that a substantial part of the national emission commitments are to be fulfilled domestically (ibid.). However, it is not clear how this will be implemented. The Government has so far not indicated that it will put any restrictions on entities buying emission quotas from abroad. Without restrictions in this regard, the market (the price mechanism) will determine how large a part of Norway's emission commitments will be carried out abroad. Recent developments in the international politics of climate change make it likely that the international quota price will be low. There are at least three important reasons for this. First, several countries in Eastern Europe, in particular Russia, have a substantial store of emission quotas to sell without having to implement corresponding domestic measures. This phenomenon, known as 'hot air', is a result of the closure of many fossil-fuel intensive factories in the wake of the breakdown of communism.[5] Second, the largest potential buyer of emission quotas, the United States, withdrew from the Kyoto Protocol in 2001. And third, a decision was made at the 7th Conference of the Parties (COP7) in Marrakech to adopt more flexible rules regarding credit for carbon sequestration.

All these factors point in the same direction: In an international market for emission quotas, the supply side is likely to be substantially larger than the demand side. If the international quota price becomes as low as USD 5 per metric ton CO_2 equivalents as some studies indicate (Hagem and Holtsmark, 2001), a substantial part of the Norwegian emission commitment is likely to be fulfilled abroad. Furthermore, there is some potential for cheap domestic emission reductions. The Norwegian Pollution Control Authority (PCA) has estimated that Norway can close 50 per cent of the gap between the baseline and emission target by implementing domestic measures that have a marginal socio-economic cost of less than NOK 200 per metric ton of CO_2 (PCA, 2000).[6]

The conclusion is that Norway can fulfil its Kyoto emission commitments at low socio-economic cost by implementing some key domestic measures within the industry combined with extensive use of the flexibility mechanisms of the Kyoto Protocol. However, if a substantial part of Norway's emission commitments are to be acheived nationally, it can be difficult for Norway to fulfil its promises. The present prime minister, Kjell Magne Bondevik from the Christian Democrats, admitted in autumn of 2003 that the Kyoto target can be out of reach for Norway (*Dagsavisen*, 2003).

Above we have analyzed the prospects for Norway fulfilling its Kyoto target. However, the Kyoto target lies several years into the future. In this study we have to analyze domestic goal attainment in relation to what has been achieved so far compared to what would have occurred in the absence of a national climate policy. The main policy instrument targeting CO_2 emissions for the period 1990 to 2000 has been the carbon taxes. Statistics Norway estimates that the increase in CO_2 emissions would have been 21.1 per cent in the period 1990-99 instead of the actual increase of 18.7 per cent if no carbon tax had been implemented (*Gemini*, 2002, p.18). We therefore conclude a low domestic goal attainment for Norway in the field of climate change.

Success in Attaining International Climate Goals

In the international climate negotiations, Norway has been particularly interested in four key issues. First of all Norway has been arguing for an international climate regime with *binding emission commitments* for the industrialized countries. The negotiations on an international treaty on climate change started early in 1991. After less than two years of negotiations there was an agreement on establishing the United Nations Framework Convention on Climate Change (UNFCCC). It entered into force in March 1994 when the 50th country (Portugal) ratified the treaty. However, while the UNFCCC was strong on building an institutional framework for further international cooperation in the field of climate change, it was weak on emission commitments. The climate convention recommended only that the industrialized countries stabilize their emissions of carbon dioxide by year 2000 at 1990 levels (Lundli, 1996).

Norway did not succeed in its efforts to establish a climate treaty with binding emission commitments at this time. However, Norway was still somewhat satisfied with the climate convention that was agreed upon in Rio. The UNFCCC constructed an institutional framework for future negotiations on binding emission commitments. It also instructed countries to develop national strategies to reduce emissions of greenhouse gases. Furthermore, the industrialized members of this new environmental regime had to produce national emission inventories for greenhouse gases. A main objective of one of the sub entities of the climate convention, the Subsidiary Body on Scientific and Technological Advice (SBSTA), was to develop and agree upon a common method to calculate such emissions.

The institutional framework for further international cooperation in the field of climate change finally produced the result Norway was hoping for. At the Third Conference of the Parties (COP3) in Kyoto, Japan, a Protocol with binding emission commitments for industrialized countries was agreed upon. Because at least 55 countries must ratify the treaty, including industrialized countries that together represent more than 55 per cent of the total emissions of industrialized countries, the withdrawal of the United States from the Kyoto Protocol in 2001 made the prospects for the treaty entering into force bleak. However, intense diplomatic efforts from the European Union saved the Protocol for the time being (Ministry of the Environment, 2002). If Russia ratifies the Kyoto Protocol, it can enter into force. As mentioned previously, it is still an open question whether Russia will ratify the treaty or not.

The second aspect of particular interest to Norway during the negotiations was the question of differentiating emission commitments between the industrialized countries. Norway wanted a Protocol that takes into consideration the different national contexts, arguing that it would be more costly for Norway to reduce its emissions of greenhouse gases than for many other industrialized countries. This is because electricity production in Norway is fully based on hydropower, Norway has an expanding oil- and gas sector, and that a decentralized settlement pattern makes it difficult to reduce emissions from the transport sector (Langhelle, 2000, pp.191-2). The Norwegian position on differentiated emission commitments had little chance of success at the outset. Traditionally, emission commitments have been across the board, with equal per centage reduction targets for all countries signing the treaty (Underdal, 2001). Furthermore, in the international climate negotiations both the European Union and the United States hesitated to accept the arguments for differentiated emission commitments. However, together with Japan, Australia and Russia, Norway finally managed to pursue its view on the international regime (Ministry of the Environment, 2002). This is why Norway was allowed to increase its emissions by one per cent while the Kyoto Protocol requires global emissions to be reduced by 5.2 per cent in total. While the Kyoto Protocol commits the industrialized countries to reducing their greenhouse gas emissions by 5.2 per cent in total, Norway is allowed to increase its emissions by 1 per cent between 1990 and 2010.

A third important issue for Norway in the international climate negotiations was the type of greenhouse gases to be included in the treaty. Norway argued for a treaty that would include six different greenhouse gases or groups of gases. The non-binding stabilization target in the climate convention referred to only one particular greenhouse gas, CO_2. When the negotiations on a Protocol with binding emission targets began in 1995, Norway argued strongly for a Protocol that included additional greenhouse gases – specifically CH_4, N_2O, HFC, SF_6 and PFCs. It would make a large difference to Norway's abatement costs whether an emission target referred only to CO_2 or to a basket of different greenhouse gases. The Norwegian emissions of CO_2 were expected to increase substantially after 1990 (the reference year in the climate treaty), while the emissions of the other five gases could more easily be curbed. For example, the aluminium industry in Norway reduced its emissions of PFCs substantially in the years after 1990 by implementing technological modifications. Again the final design of the climate regime was in line with Norway's negotiation positions. The binding emission target in the Kyoto Protocol referred exactly to the six different types of greenhouse gases listed above.

Fourth, Norway supported an international climate treaty that would allow emission trading and joint implementation between countries, arguing that such mechanisms would make it possible to achieve cost effectiveness across countries. This would make it possible to obtain larger global emission reductions than what would be the case with a treaty that only allowed domestic implementation. In fact, it was Norway that originally introduced the idea of joint implementation. The climate convention embraced the idea of flexibility mechanisms in general. It also included a pilot phase on joint implementation for the period 1995-2000 in order to gain experience with this mechanism. The Kyoto Protocol that was adopted in 1997 states that a country can fulfil its emission commitments partly by using three different flexibility mechanisms. During the negotiations prior to and after the signing of the Kyoto Protocol, the European Union wanted a cap on the use of these mechanisms. Norway opposed this view. At COP7 in Marrakech the European Union decided to withdraw this proposal in order to save the Protocol (Ministry of the Environment, 2002).

The main conclusion to be drawn so far is that Norway has been successful in influencing the international climate regime. Although it argued for a more ambitious emission target for the industrialized countries, the overall conclusion is still that the design of the regime to a considerable degree overlaps with the Norwegian negotiation positions. At the same time, the domestic achievements with regard to reductions in emissions of carbon dioxide have not been impressive.

Explaining Norway's Goal Attainment

The question before us now is why Norway was able to achieve such a high level of goal attainment at the international level but not at the domestic level. First, we look at factors at both the national and international level that can explain Norway's goal attainment in the international climate negotiations. In the next section we try to identify the key reasons that can explain Norway's implementation record and implementation prospects at the domestic level in the field of climate change. Thereafter we investigate to what degree a lack of coordination between the external and internal dimension of Norwegian climate policies can have affected the Norwegian goal attainment. Finally, we discuss the importance of problem type in understanding the Norwegian goal attainment in the field of climate change.

Explaining Goal Attainment in Regime Design

A national consensus based on international cost-effectiveness. The architect behind a cost-effectiveness strategy in the international climate negotiations was the Ministry of Finance (Skjærseth and Rosendal, 1995, p.166).[7] Cost-effectiveness means that (global) environmental targets are to be met at the lowest possible socio-economic cost. This implies that the cheapest emission measures should be implemented first, that the emission targets should be differentiated between countries in accordance with the different national abatement cost structures, that as many greenhouse gases as possible should be included in an international climate treaty, and that carbon sequestration measures should be allowed. Perhaps most importantly, it implies that the international climate treaty should include flexibility mechanisms such as emission trading and joint implementation in order to further improve the cost effectiveness between countries.

Reitan (1998, pp.178-81) draws on two factors when she explains why Norway chose to focus on international cost effectiveness. The first refers to a transition in the underlying economic theory upon which the Ministry of Finance bases its policy recommendations – from 'Keynesism' in the 1970s to neo-liberal economic theory in the 1990s. The second is that an international climate treaty designed to be cost-effective would simply imply lower socio-economic costs for Norway.

As mentioned, the Norwegian Government has been speaking with one voice 'upstream', despite a large number of ministries being involved in formulating the Norwegian negotiation positions. The overall responsibility for coordinating the national policies and for heading the negotiation delegation is shared between the Ministry of Foreign Affairs and the Ministry of the Environment (Skjærseth and Rosendal, 1995, p.167). However, the ministries of finance, oil and energy, and industry and trade have also played a key role in formulating the Norwegian climate positions. Furthermore, the Prime Minister's Office, the Ministry of Agriculture and the Ministry of Transport and Communications are also to some degree players in this game.

However, one important difference in opinion between the ministries can be observed. The Ministry of the Environment views the flexibility mechanisms more critically than the other ministries, and emphasizes to a larger degree the importance of designing the climate regime in such a way that the problem of 'hot air' is reduced. At the other end of the spectrum we find the Ministry of Finance, which has been the strongest supporter of getting flexibility mechanisms integrated in the international climate treaty. We can see this difference in attitude towards the flexibility mechanisms partly as a result of different professions. The engineers (the environmental authorities) are more optimistic about the potential to reduce domestic emissions of greenhouse gases at a low cost (and therefore anticipate less need for emission trading). The economists (the Ministry of Finance) believe that the carbon tax has already released the cheapest reduction measures in Norway and that further reductions would imply high costs.[8] Thus they emphasize that extended use of the flexibility mechanisms would imply both increased global reductions in greenhouse gas emissions and reduced costs for Norway.

The national consensus on negotiation strategy was not based on international cost effectiveness alone, but international cost effectiveness *combined with an ambitious overall emission target for the industrialized countries*. This formula made it possible for Norway to combine its material and ideal interests. Furthermore, Norway had learned from the acid rain and ozone cases that taking ambitious positions in the negotiations (being a forerunner) could accelerate the international negotiation processes and have the effect of 'shaming states into agreeing to emission cuts because other states have publicly accepted them' (DeSombre, 2002, p.178).

We have so far looked at how domestic developments have influenced Norway's ability to influence the regime outcome. Now we move on to consider developments at the international level that have made it easier for Norway to get the treaty it wanted.

International focus on cost effectiveness and processes for learning. Developments within other environmental regimes made it easier for Norway to gain international acceptance for its positions in the climate negotiations. The historical practice of having across-the-board emission targets (measured in per centage change) for all developed countries began to erode. For example, the second sulphur protocol under the CLRTAP-regime (signed in 1994) accepted the principle of differentiated emission commitments (UNECE, 2003). This meant that the neo-liberal focus on incentive structures and cost effectiveness gained ground internationally.[9]

Knowledge is important in the issue of anthropogenic climate change. States tend to use scientific uncertainty as an excuse for not taking action on environmental problems (DeSombre, 2002). Institutionalizing a process for evaluating the scientific knowledge had proven to be valuable in the ozone case. Intergovernmental procedures for reaching a transatlantic scientific consensus were a precondition for developing an effective international ozone regime (Lundli, 1996). So much so that it became a model for the climate regime. The establishment of the Intergovernmental Panel of Climate Change (IPCC) in 1988 was a successful move in the effort to launch a process for establishing a global consensus on *the science of climate change.*[10] However, the international climate regime also institutionalized a process for gaining experience and knowledge about the flexibility mechanisms through the establishment of the Subsidiary Body on Scientific and Technological Advice (SBSTA). SBSTA has been an arena for sharing the experiences gained in the pilot phase on joint implementation that ran from 1995 to 2000. It has also been an arena for discussing different ways to implement the flexibility mechanisms and build a regulatory framework. The way that the climate regime has institutionalized a learning process in this field has been important in obtaining global acceptance for these policy instruments. The only former experience with emission trading is the emission trading system for sulphur dioxide (SO_2) that was introduced in the United States in 1990.[11]

To develop an effective international climate regime, developing countries must accept binding emission commitments at some point in time. The experiences gained in the ozone case have been valuable in this regard. In the ozone regime, a financial mechanism (the Multilateral Fund) was established to assist the developing countries in their phase-out of ozone-depleting gases. Until this mechanism was in place, key developing countries such as Brazil, India and China refused to take action against national production and consumption of ozone-depleting gases

(DeSombre, 2002, p.112-13). In the climate negotiations, Norway has played a role as a mediator between the developed and developing countries – supporting the establishment of similar financial mechanisms in the climate regime.

When explaining why Norway managed to get the treaty it wanted, we also have to pay attention to two important characteristics of the international climate regime. First of all, the international climate regime embraces nearly all countries in the world.[12] And second, the decision rules are based on the principle of consensus. According to classical theory on international negotiations, the least enthusiastic countries determine the overall level of ambition for the regime, and thus to a large extent its design (Underdal, 2001). This has also been the case in the climate regime. The European Union is often characterized as the most 'progressive' party in the international climate negotiations. If it had been up to the European Union itself to decide, the design of the climate regime would not have satisfied Norway. The EU argued for equal emission reductions targets for all industrialized countries. Furthermore, it wanted a climate treaty that included only three specific greenhouse gases (CO_2, CH_4 and N_2O). And finally, the EU argued strongly against the use of flexibility mechanisms. Fortunately for Norway, important countries such as Japan, Australia and Russia supported a treaty with differentiated emission commitments. The United States, Australia and Russia wanted the treaty to include six different greenhouse gases and the option to use flexibility mechanisms. However, the fact that the least enthusiastic countries decided how the climate regime was to be designed also had implications for the overall emission target for the industrialized countries. As mentioned, Norway's and the EU's proposal for a 10 to 15 per cent emission target for these countries did not materialize.

We have so far identified some key factors in understanding why the international climate regime outcome correlated to a high degree with the Norwegian negotiation positions. Now we will try to explain the Norwegian achievements with regard to domestic implementation.

Explaining Domestic Implementation

The climate issue received a lot of attention from the political parties during the 1990s. This was particularly the case in 1989 – when the stabilization target was adopted. Later in the 1990s the question of building gas-fired power plants was fiercely debated in the Parliament. In fact a minority government decided in 2000 to step down over this matter. However, in our analysis of the Norwegian implementation record we will more or less omit the political variable. There are two reasons for this. First of all we would argue that despite shifting minority Governments throughout the 1990s, the national implementation policy has not changed to any substantial degree. This is due to an existing *majority consensus* in the Parliament with regard to the main lines in the national climate policy. This majority has consisted of the Labour Party and the Conservatives. For example, this majority has always seen the unilateral stabilization target primarily as a way to push for international cooperation. The opposition consisting of the Socialist

Left, the Centre Party, the Christian Democrats and the Liberal Party has not succeeded with its arguments against gas-fired power plants and its proposal to extend the carbon tax.[13]

Second, we would argue in line with Reitan and Stigen (2000, pp.416-17) that since 1992 national climate policy formation has largely taken place in closed negotiations between the various ministries. Due to the complexity of the climate issue, the formulation of policy strategies has to some degree been left to the governmental experts (bureaucrats). A possible example of bureaucrats persuading policymakers was observed in 1997 when the centre-minority Government gradually modified its view regarding the flexibility mechanisms. Before entering into office, the Christian Democrats, the Centre Party and the Liberal Party all expressed scepticism to the flexibility mechanisms. However, soon after entering into office, this Government worked enthusiastically for a protocol including these mechanisms, and it expressed satisfaction with the outcome in Kyoto in this regard.[14]

We will explain the domestic implementation processes in light of a struggle between different ministries and society sectors. Furthermore, we believe that knowledge is important in the climate issue due to the complexity of this field. This means that researchers and governmental experts have an advantage over politicians when it comes to agenda setting and suggesting policy strategies. In addition to focusing on variables at the national level we will also look at how developments at the international (regime) level have influenced domestic implementation.

Domestic institutions: From conflict to consensus on emission trading systems. While there has been a broad consensus between the ministries regarding what positions Norway ought to take in the international climate negotiations, this has not been the case when it comes to which climate policies to implement nationally. In general the conflicts between the ministries can be related to two interconnected issues: taking on a unilateral stabilization target before a binding international treaty is agreed upon, and exempting the process industry from the unilateral carbon tax. The lines of conflict between the different ministries have varied according to which of these two issues were being discussed. However, the conflicts between the ministries are considerably lower at present than what was the case in the 1990s. The reduced level of conflict is due to the prospects of an international treaty with binding emission commitments entering into force and the fact that the Parliament has decided to replace the controversial carbon tax with a national system for emission trading.

When the unilateral emission target was adopted in 1989, there was broad national consensus. This was also the case when carbon taxes were introduced as the most important policy instrument to curb emissions of CO_2, and they were supposed to be differentiated according to the carbon intensity of the different fuels. However, this consensus soon disintegrated. An international climate treaty with binding emission commitments failed to materialize. When the Environmental Tax Committee (*Miljøavgiftsutvalget*) presented its final conclusions in February 1992, a majority of its members recommended a rejection of the unilateral stabilization target. The Ministry of the Environment, on the other hand, objected to the Committee expressing an opinion in this matter.[15] The Ministry of the Environment became the strongest supporter of Norway fulfilling its unilateral emission target despite the lack of an international treaty with binding emission commitments (Reitan, 1998, p.139).

The strongest opponent of the Ministry of the Environment in this matter was the Ministry of Finance. Using macro-economic models, it argued that a unilateral policy to fulfil the stabilization target would considerably reduce GNP and substantially increase the unemployment rate (*Dagens Næringsliv*, 1992). Other key ministries such as the Ministry of Trade and Industry expressed similar worries. It was also argued that unilateral policy could backfire and global emissions of greenhouse gases actually increase if the relatively 'clean' Norwegian process industry had to scale down its production due to a unilateral climate policy (Reitan, 1998, p.178).[16] The majority of members in the Environmental Tax Committee emphasized that it was more important for Norway to work for a cost-effective international climate treaty than to play the role as a forerunner in this field (Reitan, 1998, p.138). It was necessary to have 'a flexible attitude to when a stabilization in the emissions of carbon dioxide can take place' (Ministry of Finance, 1992, p.17).

The Ministry of the Environment was virtually the only ministry that supported meeting the stabilization target despite the absence of an international treaty with binding emission commitments. The majority of the other ministries had from the outset seen the stabilization target only as a means to influence international action. Previously we have seen a similar majority in the Parliament with regard to the interpretation of the stabilization target.[17]

However, when it came to the question of extending the carbon tax to all emission sources, there was a common understanding between the Ministry of the Environment, the Ministry of Finance and the Ministry of Transport and Communications. The first step in introducing a carbon tax in Norway was taken in 1991, when it was introduced for mineral oil, petrol and for fossil fuel consumption on the continental shelf. A further increase in the number of emission sources to be taxed as well as an increase in the tax level was anticipated. The Green Tax Commission (*Grønn skattekommisjon*) was appointed in 1994 to make recommendations for how the carbon tax could be developed over time in Norway (Reitan, 1998). The work of the Green Tax Commission revealed that the Ministry of Trade and Industry and the Ministry of Oil and Energy did not support an equal carbon tax for all emission sources. Together with the industry itself and a majority in the Parliament, the Commission argued that a tax exemption for the process industry was necessary from both an industrial and an environmental perspective (Reitan, 1998).

The question of extending the carbon tax to all emission sources was also linked to the question of building the first power plants based on natural gas in Norway. If the gas-fired power plants had to pay a tax for their emissions of CO_2, they would most probably not be realized. The issue of gas-fired power plants became the single most debated environmental issue in Norway during the 1990s. As already mentioned, a minority centre coalition Government stepped down over this issue in March 2000. It could not accept that the majority in Parliament instructed the Government to reverse a decision taken by the Pollution Control Authority on this matter. Based on the Pollution Control Act, the Pollution Control Authority had imposed very strict emission regulations for two planned gas-fired power plants. A majority in the Parliament, on the other hand, did not want to apply stricter emission permits to such types of power plants than what was the case in other European countries (Ministry of the Environment, 2000).

The conflict between the various ministries resulted in a stalemate with regard to implementation. A majority of the ministries and policymakers opposed further national efforts in the absence of an international treaty with binding emissions targets. When (and if) the Kyoto Protocol enters into force, this hindrance to further national climate efforts will be removed. Furthermore, in spring 1998 the Parliament settled the issue regarding the question of extending the carbon tax to other emission sources. A majority in the Parliament not only voted down the extension proposal – it also decided to consider replacing the carbon tax with a national emission permit system (Ministry of the Environment, 1998; Kaasa and Malvik, 2000). In spring 2002 the Parliament backed a government proposal to fully implement such a policy instrument from 2008 at the latest.

The introduction of an emission permit system has considerably reduced the level of conflict in Norwegian climate change politics. The various political parties and ministries have put the conflicts about the carbon tax behind them and are instead focusing on how to implement the national emission trading system. Industry representatives are also satisfied with the transition to the new system, not least because it was industry representatives who originally proposed the system, and its adoption shows that industry indeed has a voice in climate policy formation. (Kasa and Malvik, 2000).

The transition from carbon taxes to emission permits will most likely imply lower socio-economic costs because the international quota price is expected to be considerably lower than the present cost of carbon taxes in Norway. Based on present knowledge, the only domestic target group that would face 'worse' conditions is the process industry, i.e. those CO_2 sources that have been exempted from carbon taxation so far. This is also reflected in the present political climate; there is in general no longer much criticism of the national climate policies by private actors, except the process industry.

One important reason for switching to a national emission permit system is that experts believe this policy instrument will better enable Norway to meet its Kyoto emission target (Ministry of the Environment, 2002). As we have seen previously, the carbon tax has only marginally reduced the growth in emission of CO_2.

International environmental regimes and domestic implementation. To understand the implementation processes in Norway, it is also necessary to take a closer look at some important developments at the international level. Developments within the core regime, that is the climate regime itself as well as developments within other linked regimes, have influenced the Norwegian implementation strategy.

The core regime – building trust between parties. A 'strong' climate regime is more likely to facilitate the implementation of domestic emission reduction measures than a 'weak' regime. Regime strength is in our analysis seen as a function of (1) legal status of commitments; (2) specificity of commitments; (3) verification of compliance; and (4) enforcement procedures (as outlined in Chapter 2). We would argue that the strength of the climate regime has increased over time. When the UNFCCC entered into force in 1994, it was weak when it came to the question of emission commitments. It only recommended that industrialized countries stabilize their emissions of CO_2 in 2000 at 1990 levels. This was a non-binding recommendation, or differently stated, soft law. If the Kyoto Protocol enters into force, on the other hand, legally binding emission commitments will be in place.

The climate regime has always been strong on the administrative side. From its inception, the UNFCCC has required member states to develop national greenhouse gas emission inventories and to submit these data (as well as a national report) on a regular basis to the secretariat. It also introduced a common methodology for reporting national emissions. These processes have largely discouraged countries from cheating when it comes to fulfilling emission commitments. The UNFCCC also decided to let intergovernmental expert groups visit and evaluate different aspects of the emission inventories of member states from time to time.

The specificity of commitments in the climate regime is high. The emission commitments given by the Kyoto Protocol are clearly specified in numbers for each industrialized country. Furthermore, the target years (2008-12) and reference year are clearly defined in the treaty. And the common methodology to calculate and report national emissions of greenhouse gases developed under the UNFCC makes it relatively easy to verify whether the parties achieve their emission commitments or not. Hence, the high specificity of commitments makes it possible to verify compliance. However, verification of compliance can still be problematic in two important areas. First of all, the emission inventories for the east European countries and Russia are of low quality. If emission trading and joint implementation between Annex I countries is to make sense, the quality of these inventories must be improved substantially. And second, greater knowledge is needed of how carbon sequestration measures (e.g., reforestation and other 'sink' options) affect the total carbon budget. At present there is a considerable scientific controversy over how to estimate the carbon effect of different types of sink measures.

At COP7 in Marrakech, enforcement procedures were established to make defection more costly. Industrialized parties that do not fulfil their emission commitments face restrictions on future use of the flexibility mechanisms and tighter future national emission commitments than otherwise.

Linked regimes affecting domestic implementation. Norway's membership in the European Economic Area (EEA) implies that it is more or less fully integrated into EU environmental policy. There is one particular EU directive that can affect the Norwegian climate policies to a considerable degree: the EU Directive on emission trading. This directive was adopted in July 2003 and will enter into force on 1 January 2005. Because of its membership in the EEA, Norway normally has to implement all EU directives in the field of environmental policy. However, in the case of the directive on emission trading Norway is most probably not obliged to adopt it into national policy. Norway is not part of the EU 'emission bubble', and it is therefore not likely that it has to follow the main lines in the EU climate policy on implementation. On the other hand, it can be difficult from a political perspective to stand on the outside of this directive. Norway also wants to connect its future emission trading scheme to the EU emission trading scheme, but this presupposes that the design of the EU emission trading system suits Norwegian interests. The authorities are planning to present the details of a national emission trading scheme during the first half of 2004. At this time, the authorities will make up their mind about connecting to the EU emission system or not. Let us take a closer look at the main aspects of the EU Directive on Emission Trading.

The EU Directive on emission trading is for the period 2005-07 limited to large energy production entities (power plants based on fossil fuels), refineries, steel producers and the pulp and paper industry. At the outset, member countries are only allowed to include other types of entities from 2008 at the earliest. However, in 2004 and 2006, the EU will evaluate the emission trading system to decide whether other sectors such as the chemical industry, aluminium production or transport are to be included. The EU emission trading system as it is designed today does not mesh well with the corresponding system that is to be implemented in Norway. In Norway the government has proposed that an early emission trading system limited to the process industry be operational from 2005. This can be problematic if Norway adopts the EU emission trading directive, since the directive does not include the process industry.

A main principle in the EU directive is that most of the emission quotas are to be distributed to the entities *free of charge*. From 2005, only five per cent of the emission quotas (measured in CO_2 equivalents) are to be sold, while from 2008 this proportion is increased to ten per cent. The Norwegian government, on the other hand, has proposed that most of the emission certificates be sold or auctioned to those entities that are to be covered by the system (Ministry of the Environment, 2002). One reason for this is that the carbon tax is to be removed at the same time as the national emission quota system is introduced. If the emission certificates are distributed free of charge, then those entities that today pay the carbon tax (such as the transport sector and oil- and gas producers) would achieve a considerable reduction in their 'climate costs'.

Thus the EU Directive on emission trading will affect domestic implementation of climate policy in Norway to a considerable degree – if the directive is adopted by Norway. Due to the mismatch between the EU and Norwegian emission trading systems, is not likely that Norway will adopt the directive in the next few years. However, if the EU emission directive is modified over time so that it takes into account the Norwegian priorities, then it is likely that the directive will become Norwegian law.

The goals of other environmental regimes generally do not hinder Norway from reaching its Kyoto target, as can be seen through the examples of the ozone regime, the 'acid rain' regime (CLRTAP) and the biodiversity regime. The *international ozone regime* is to a large degree coordinated with the climate regime. For example, gases that are included in the Montreal Protocol and its amendments are excluded from the climate regime. However, the members of the international ozone regime partially fulfil their obligations by substituting chlorofluorocarbons (CFCs) with the greenhouse gas hydrofluorcarbons (HFCs). HFCs are greenhouse gases with high global warming potentials (PCA, 2000). Hence, the international effort to repair the ozone layer has to a minor degree accelerated the problem of anthropogenic climate change. However, it is expected that it will be technologically possible to develop new substitutes at low costs in the near future – in other words, that the problem of HFC emissions will be a temporary one (PCA, 2000).

The *Convention on Long-Range Transboundary Air Pollution* (CLRTAP) with its affiliated protocols seeks to reduce the emissions of SO_2, NO_x, ammonia gas (NH_3), and non-methane volatile organic compounds (NMVOCs) (DeSombre, 2002). Measures taken to reduce emissions of these substances also affect energy efficiencies in various ways, and consequently, emissions of CO_2. Reductions in the emissions of NO_x are often followed by increased fuel consumption. On the other hand, reductions in the emissions of NMVOC also contribute to lower emissions of CO_2.[18] In an evaluation of possible domestic climate policy measures carried out by the Pollution Control Authority, measures to reduce the emissions of NMVOCs were identified as being among the most economically viable (PCA, 2000). In sum, the CLRTAP regime has a positive effect on the Norwegian goal attainment in the area of climate change.

The main objective of the *Convention on Biological Diversity* is to ensure the conservation of biological diversity and the sustainable use of its components (Bergesen and Parmann, 1995, p.184). It is debatable as to what degree large-scale climate-motivated forestation projects can be defended in relation to the goals of the biodiversity regime. However, although Norway has supported the sink option in the Kyoto Protocol, it has decided that domestic forestation projects should not be part of the national climate strategy (Ministry of the Environment, 2002). On the other hand, several of the joint implementation projects that Norway implemented in developing countries during the 1990s (as part of the pilot phase on joint implementation) were sink projects (Ministry of the Environment, 2001). It is likely that this may also be the case during the operational phase of the Kyoto Protocol. However, Norway has emphasized that the implementation of such measures must be done carefully to avoid loss of biodiversity.

We have so far looked at factors explaining the Norwegian goal attainment in the international climate negotiations and regarding domestic implementation. We will now investigate to what degree a lack of coordination between the external and internal dimension of Norwegian climate policies may have played a role.

Coordination of the International and Domestic Dimension

In the first formative years of Norwegian climate policy (1988-90), the Ministry of the Environment dominated (Andresen et al., 2002, p.2). The problem of climate change was an environmental problem and was believed to be the sole responsibility of the corresponding ministry. Furthermore, industry did not pay much attention to this new policy field at this time. Nor did industry representatives produce any official comments to the White Paper on the follow-up of the

Brundtland Commission, which suggested that a national emission target be adopted (Bolstad, 1993, p.24). Nor was industry involved when the unilateral stabilization target was set in 1989.

Industry entered the scene when the authorities started to evaluate what policy measures to adopt to fulfil the national stabilization target. Industry did not mobilize against the introduction of a carbon tax on petrol and mineral oil because it did not affect export-oriented activities to any substantial degree. However, when the Government began to consider extending the carbon tax to also include emissions from the continental shelf and the process industry, a massive mobilization against these proposals was triggered. Within the Government, a similar mobilization took place (the Ministry of Trade and Industry, and the Ministry of Oil and Energy). Hence, the Ministry of the Environment was no longer the dominant player in the game (Reitan, 1998).

Since 1991, all relevant ministries and domestic target groups have been actively involved in shaping the national climate change policy. This regards both the international and domestic dimension of the climate policy – what positions to take in the international climate negotiations and what implementation steps to take nationally. If the affected ministries and target groups had been more actively involved *before* 1991, the adoption of the stabilization target would have been less likely. A target would probably have been adopted also in such a case, but it would most probably have been a target more in line with the original proposal from the Labour Government at that time (stabilizing CO_2 emissions by 2000). The industry could have used its historically close contacts with the Conservatives and the Labour party to secure a majority in Parliament for such an emission target.

Although the affected ministries and target groups have been important players in the climate policy field since 1991, the Government knew at the time of the adoption that the unilateral stabilization target could be difficult to achieve. The so-called SIMEN report that Statistics Norway developed in 1989 at the request of the authorities emphasized that the effects on the economy of the use of carbon taxes would largely depend on whether this was a unilateral policy or a policy that went hand-in-hand with a similar use of carbon taxes in other industrialized countries (SIMEN, 1989). A unilateral policy would in particular affect the export-oriented process industry, according to this report. Despite these results, the Government and in particular the Ministry of the Environment wanted to establish this ambitious emission target to be a forerunner in this new environmental policy field. If the other ministries had been more active in these formative years of a national climate change policy, the role of environmental forerunner would most probably have been played down. It would also have made it less embarrassing for Norway with regard to accomplishing national emission goals.

Problem Type as an Alternative Explanatory Approach

The climate issue is an extremely malign problem – both at the international and the national levels. At the international level, there is an asymmetrical distribution of vulnerability, abatement costs and capacity to solve the problem. The industrialized countries in the northern hemisphere, which have created the problem in the first place, lack the crucial incentive of great vulnerability to actually mobilize their resources in a struggle against climate change. On the other hand, the countries that are expected to be the most affected by anthropogenic climate change, the developing countries, do not have the capital or technology needed to reduce their own emissions. In addition, the emission of greenhouse gases per capita is many times higher in the industrialized countries than in the developing countries. The contrast to the successful international ozone regime is remarkable. In the case of ozone depletion, those countries responsible for the emissions of ozone depleting gases were those most vulnerable to the effects flowing from the problem; the costs of reducing the emissions were low and continually falling; and the responsible countries were highly developed economically and technologically, making them able to abate the emissions (Lundli, 1996). The fact that the climate issue is an extremely malign problem makes it difficult for Norway (and the EU) to gain international acceptance for ambitious emission targets for the industrialized countries. However, the United States has been the dominant actor in the international negotiations. The compatibility between the US and Norwegian positions ensured a high match between the Protocol and Norwegian positions.

At the national level the benefits of an improved global climate is questionable from a narrow point of view. Although more rain and flooding can be expected, an increase in the mean temperature can also have some positive consequences for Norway (for example improved agricultural yield; less need for heating). Furthermore, the benefits are several decades into the future. The costs related to a fulfilment of the national Kyoto target are concentrated to the processing industry. Implementing an effective climate policy is difficult under such circumstances. Few domestic actors (if any) besides the environmental organizations are likely to argue for an extensive use of policy instruments that impose costs on industry or consumers. On the other hand, a strong opposition against an ambitious national climate policy is expected to continue from parts of the industry, the processing industry in particular.

Conclusion

In 2002, Norway was the third largest exporter of crude oil in the world. It is also one of the countries with the highest climate abatement costs. At the same time, Norwegian vulnerability to anthropogenic climate change is believed to be relatively low. An interest-based approach to foreign policy would predict that Norway would play the role of laggard in the international negotiations on climate change.

However, Norway has since the release of the report *Our Common Future* in 1987 explicitly wanted to play a role as a forerunner in environmental matters (Skjærseth and Rosendal, 1995, p.163). In the climate issue, the Ministry of Finance found the key to Norway playing the role of forerunner without weakening the competitiveness of Norwegian industry. The formula was *international cost-effectiveness* combined with *an ambitious overall emission target for the industrialized countries*. All ministries backed this formula. The strategy of international cost-effectiveness included differentiation of targets, flexibility mechanisms, and a comprehensive approach to emissions reduction that allowed reduction of six different types of greenhouse gases. The Kyoto Protocol included exactly these elements. One important reason for this outcome was that the Norwegian focus on international cost-effectiveness coincided with the interests of some key (laggard) countries. Furthermore, market-based policy instruments gained popularity in most countries in the 1990s. Other international environmental regimes began to include differentiated commitments between countries during the 1990s, paving the way for a similar approach in the climate issue.

At the international level, Norway was successful in helping to direct the focus to cost-effectiveness in the climate regime. However, it did not succeed in achieving its ambitious emission reduction level. In the negotiations, Norway argued for a 10 to 15 per cent reduction in the greenhouse gas emissions for the industrialized countries in total. The corresponding outcome in Kyoto was only a 5.2 per cent reduction target.

Emissions of greenhouse gases have increased rapidly in Norway since 1990. While Norway has largely succeeded in attaining its goals internationally, it has not managed to substantially reduce the growth in domestic emissions compared to a business-as-usual development. Despite rapidly increasing domestic emissions of greenhouse gases, Norway can use the flexibility mechanisms of the Kyoto Protocol to meet its Kyoto target at low socio-economic costs. Two important developments at the international level have considerably reduced Norway's expected costs associated with the Protocol: the withdrawal of the United States, which lowers quota prices, and the extension of the sink option. Instead of having an international climate regime that 'forces' Norway to develop and implement more ambitious climate policies, the opposite is the case at present. It is not necessary to change the behaviour of the domestic target groups to any substantial degree in order to fulfil the national Kyoto target. The target can be met by implementing measures in the process industry and by using the mechanisms of emission trading, joint implementation and the Clean Development Mechanism.

We have also seen that the adoption of an EU Directive on emission trading could further reduce the climate costs for Norwegian industry and consumers – if Norway adopts the directive in question. In this case Norway can fulfil its Kyoto target at a lower socio-economic cost per year than the costs associated with the present carbon tax. The EU Directive on emission trading which is to enter into force in 2005 implies that most of the emission quotas are to be distributed to the different entities for free. Those entities and society sectors that today are covered by the carbon tax (for example transport and the production of oil and gas), can receive most of the emission quotas needed for free in the future. It will be a challenge for the Norwegian authorities to defend its self-declared role as a forerunner in international politics of climate change if the scenario sketched above is realized. One possible option for solving such a dilemma could be to adopt a stricter national emission target than the prevailing Kyoto target. Sweden has recently taken a similar move, exactly because it believes that the present national target is too easy to comply with. By following the Swedish example, it is possible to maintain the present level of 'climate pressure' towards Norwegian industry and consumers.

Notes

1 An earlier version of this chapter was presented at the 11th Annual Norwegian Conference in Political Science, Trondheim, 8-10 January 2003. The authors are grateful for comments received from participants at this conference.

2 National emissions of greenhouse gases can fluctuate considerably from one year to another, due to particular national circumstances. Using the average emissions for five years (2008-12) prevents such particular circumstances from determining whether a country manages to comply with its climate target or not.

3 The flexibility mechanisms are designed to allow the Parties flexibility in meeting their targets. Emissions trading, as will be explained later in this chapter, allows Parties to buy and sell their emissions quotas. Joint implementation allows one industrialized country to receive credit for implementing emissions reductions in a second industrialized country. And the Clean Development Mechanism allows an industrialized country to receive credit for undertaking abatement projects in developing countries.

4 Norwegian Labour politician Gro Harlem Brundtland served as leader of the UN World Commission on Environment and Development, often called the Brundtland Commission, which published its report *Our Common Future* in 1987. She also served as Norway's prime minister in 1981, from 1986-89 and from 1990-96.

5 'Hot air' refers to a situation where emission trading between countries does not lead to real emission reductions. It is likely that Russia and some other east European countries will meet their Kyoto targets with good margin even with a business-as-usual development. This means that these countries can sell emission quotas to other countries without having to implement corresponding measures domestically.

6 In comparison, the highest level of the Carbon tax in Norway (petrol) corresponds to 300 NOK/metric ton CO_2 (Ministry of the Environment, 2001).

7 As will be explained later, there has been a dispute between the ministries regarding how the principle of cost effectiveness is to be applied *domestically*.

8 The argument is not applicable to those emission sources that have been exempted from the carbon tax. As will be explained later, the Ministry of Finance has argued against exempting the process industry from the carbon tax.

9 However, the different national emission targets with regard to the second sulphur protocol reflected first of all what emission reduction levels were *politically feasible* for the different countries (and to a lesser extent the different national abatement cost structures).

10 The IPCC is, however, not formally part of the climate convention. It is an intergovernmental organization under the World Meteorological Organization (WMO) and the United Nations Environmental Programme (UNEP).

11 This US policy instrument was implemented in response to domestic, regional (North America) and international attention (the CLRTAP regime) to the problem of acid rain (DeSombre, 2002). The fact that a national permit system for SO_2 had existed for some time in the United States made it easier for Norway to get international acceptance for including flexibility mechanisms in an international climate treaty.

12 More than 170 countries have ratified the climate convention.

[13] The Progress Party has also opposed the climate policy that has been in operation, but with an opposite sign. This right-wing party has questioned the existence of anthropogenic climate change and has therefore argued against the need for a national climate policy.

[14] An alternative explanation is that the minority government had to take into consideration that a majority in the parliament supported the flexibility mechanisms.

[15] An evaluation of the unilateral emission target was not part of the mandate of the Committee. The mandate for the Environmental Tax Committee was to consider the future use of environmental taxes.

[16] The Norwegian process industry uses carbon-free hydropower as its energy source.

[17] Different interpretations of the Norwegian stabilization target are given in the political science literature. Our argument is, however, that *a stable majority* in Parliament and among the ministries always has interpreted the target in an international context.

[18] An important source of NMVOC emissions is the loading of oil at the oil fields offshore. Norway has considerable emissions of NMVOC due to the large oil production taking place on the continental shelf.

References

Andresen, S., Kolshus, H.H. and Torvanger, A. (2002), *The Feasibility of Ambitious Climate Agreements. Norway as an Early Test Case*, Working Paper 2002:3, Center for International Climate and Environmental Research (CICERO), Oslo.

Bergesen, H.O. and Parmann, G. (eds) (1995), *Green Globe Yearbook of International Co-operation on Environment and Development*, Oxford University Press, Oxford.

Bolstad, G. (1993), *Inn i drivhuset*, Cappelen, Oslo.

Dagens Næringsliv (1992), 'Miljømålene for tøffe', February 6.

Dagsavisen (2003), 'Bondevik varsler brudd på klimaløfte', September 30.

DeSombre, E.R. (2002), *The Global Environment & World Politics. International Relations for the 21st Century*, Continuum, London.

Gemini (2002), 'Ti år med CO_2-avgifter', no. 4.

Hagem, C. and Holtsmark, B. (2001), 'From Small to Insignificant: Climate Impact of the Kyoto Protocol With and Without the US Policy', Note 2001-01, Center for International Climate and Environmental Research (CICERO), Oslo.

Kaasa, S. and Malvik, H. (2000), 'Makt, miljøpolitikk, organiserte industriinteresser og partistrategier: En analyse av de politiske barrierene mot en utvidelse av CO_2-avgiften i Norge', *Tidsskrift for samfunnsforskning*, vol. 43, no. 3, pp. 295-324.

Langhelle, O. (2000), 'Norway: Reluctantly Carrying the Torch', in W.M. Lafferty and J. Meadowcroft (eds), *Implementing Sustainable Development. Strategies and Initiatives in High Consumption Societies*, Oxford University Press, Oxford, pp. 174-208.

Lundli, H.-E. (1996), *The Politics of Ozone Depletion and Climate Change: Sources of Success and Failure*, Cand. Polit. thesis in Political Science, Norwegian University of Science and Technology, Trondheim.

Ministry of the Environment (1989), *Environment and Development. Programme for Norway's Follow-up of the Report of the World Commission on Environment and Development*, (Miljø og Utvikling. Norges oppfølging av Verdenskommisjonens rapport), Report No. 46 to the Storting (1988-89), Ministry of the Environment, Oslo.

Ministry of the Environment (1998), Norway's Follow-Up of the Kyoto Protocol (Norges oppfølging av Kyotoprotokollen.), Report No. 29 to the Storting (1997-98), Ministry of the Environment, Oslo.

Ministry of the Environment (2000), Letter to Naturkraft A/S from the Ministry of Environment, 6 October 2000 (Brev til Naturkraft A/S fra Miljøverndepartementet, 6. oktober 2000), retreived 17 December 2002 from http://odin.dep.no/md/norsk/aktuelt/pressesenter/bakgrunn/022021-990039/index-dok000-b-n-a.html.

Ministry of the Environment (2001), *Norway's Climate Policy* (Norsk klimapolitikk), Report No. 54 to the Storting (2000-2001), Ministry of the Environment, Oslo.

Ministry of the Environment (2002), *Supplementary Report to Report No. 54 to the Parliament (2001-2002), Norway's Climate Policy* (Tilleggmelding til St. meld. nr. 54 (2000-2001), Norsk klimapolitikk), Report No. 15 to the Storting (2001-2002), Ministry of the Environment, Oslo.

Ministry of Finance (1992), *Towards a more Cost-effective Environmental Policy in the 1990s* (Mot en mer kostnadseffektiv miljøpolitikk i 1990-årene), NOU report No. 1992:3, Ministry of Finance, Oslo.

PCA, Norwegian Pollution Control Authority (Statens forurensningstilsyn, SFT) (1997), *Climate Change. Causes, Effects and Solutions* (Klimaendringer. Årsaker, effekter og løsninger), TA- 1493/97, Pollution Control Authority, Oslo.

PCA, Norwegian Pollution Control Authority (Statens forurensningstilsyn, SFT) (2000), *Reduction of Climate-gas Emissions in Norway. An Analysis of Remedial Measures for 2010* (Reduksjon av klimagassutslipp i Norge. En tiltaksanalyse for 2010), Pollution Control Authority, Oslo.

Reitan, M. (1998), *Interesser og institusjoner i miljøpolitikken*, Dr.Polit. dissertation, University of Oslo, Department of Political Science, Oslo.

Reitan, M. and Stigen, I. (2000), 'Kamp om miljøet – fra politikk til administrasjon?', *Tidsskrift for samfunnsforskning*, vol. 43, no. 3, pp. 405-35.

SIMEN (1989), *Studier av industri, energi og miljø fram mot år 2000*, Statistisk Sentralbyrå, Oslo.

Skjærseth, J.B. and Rosendal, K.G. (1995), 'Norges miljø-utenrikspolitikk', in T.L. Knutsen, G.M. Sørbø and S. Gjerdåker (eds), *Norges utenrikspolitikk*, Cappelen Akademisk Forlag, Oslo, pp. 161-81.

Underdal, A. (2001), 'Internasjonale forhandlinger', in J. Hovi and R. Malnes (eds), *Normer og makt: Innføring i internasjonal politikk*, Abstrakt forlag, Oslo, pp. 293-317.

UNECE, United Nations Economic Commission for Europe, Environment and Human Settlements Division (2003), *The 1994 Oslo Protocol on Further Reduction of Sulphur Emissions*, retrieved 11 February 2003 from www.unece.org/env/lrtap/fsulf_h1.htm.

Chapter 8

Biodiversity: International Bungee Jump – Domestic Bungle

G. Kristin Rosendal

Introduction

The worldwide loss of biodiversity is recognised as one of today's greatest global environmental problems. While attention is primarily focused on tropical rainforests, all regions of the world, including Norway, suffer from deterioration and loss of habitats and species. Apart from the loss of aesthetic and intrinsic value, the loss of species and habitats threatens the provision of numerous goods and services, from food and medicine to local climate and water regulation. Land use change, fragmentation of habitats, pollution, and introduction of alien species represent major, human-induced threats to biodiversity. The most comprehensive response to this global threat is found in the international Convention on Biological Diversity (CBD).

This chapter looks into how Norwegian authorities handle biodiversity-related policies in the potential crossfire between international environmental agreements and sub-national actor interests. It demonstrates that there is a very high correlation between domestic and international goals, but that international goals have been much more easily attained. The next step involves an examination of why international goals have been more easily reached compared to domestic ones. From a wide range of sub-topics within the biodiversity field, the forest sector is selected for an in-depth analysis of how domestic and international institutions can shed light on the limited goal attainment. The analysis suggests that insufficient legal and economic instruments applied for biodiversity protection constitute the major explanatory factors. The chapter winds up with a discussion of alternative explanatory perspectives and argues that the shortcomings are confounded by a general lack of technological solutions to stem the pressure on species and habitats.

Main Goals of Norwegian Biodiversity Policy at Home and Abroad

When discussing goals for biological diversity, it must first be pointed out that the concept itself is a quite recent invention that first surfaced in the 1980s (Wilson, 1988). The establishment of national parks as a means of safeguarding outstanding natural environments for the future and the idea of species being threatened by (human-induced) extinction are, however, far from novel. The US has been a pioneer in this work by establishing the world's first national park, Yellowstone National Park, in 1872. The first Norwegian White Paper on a comprehensive plan for national parks was submitted in 1966 and included 15 areas. By 1988, 122 areas were under protection, adding up to about 4.5 per cent of the total land area in Norway. Awareness of human-induced species loss first occurred with the extinction of the big, flightless bird, the dodo (*Raphus cucullatus*) around 1650 (Quammen, 1996). In Norway, political action aimed at saving specific species from extinction dates back to the protection of beaver and swan in the late 1800s, followed by examples including the mountain fox in 1930, the snow owl in 1965, and protection of the wolf in 1973. Attempts to save specific species from human-induced overexploitation did not occur in Norway until the mid 1980s with efforts to regulate fishing in the Barents Sea.

Prior to the establishment of the CBD in 1992, conservation efforts may be seen in connection with a number of international conventions – mainly the Bern Convention of 1979 (on the protection of wild species of plants and animals and their habitats), the 1971 Ramsar Convention on Wetlands, the Bonn Convention of 1979 (on conservation of migratory species of animals – CMS), the Convention on Trade in Endangered Species (CITES, 1973), and various recommendations from the International Council for the Exploration of the Sea (ICES). Norway is also a party to the Convention for the Conservation of Salmon in the North Atlantic and the 1992 OSPAR Convention for the Protection of the Marine Environment of the Northeast Atlantic.

The CBD represented a change in the philosophy of nature preservation, from seeing human activity in direct opposition to nature, to integrating ideas of sustainable use and equitable sharing on an equal footing with conservation of species and habitats worldwide. It brought about recognition of the skewed geographical distribution of biodiversity and biotechnology – the former largely an asset of the tropical, often poor countries of the South, and the latter dominated by corporations in the North. In recognition of the great contributions of biodiversity to global welfare, the CBD introduced a number of mechanisms aimed at ensuring a more equitable distribution of costs and benefits among users and contributors of biodiversity's genetic resources. At the same time, the CBD brings with it the acknowledgement that local biodiversity within each country and region is equally important as the biodiversity 'hotspots' of, for instance, tropical rainforests. This implies that each country takes on both domestic and global obligations when ratifying the CBD. These ideas are very much in line with Norwegian domestic and international goals for biodiversity.

Goals Adopted: Late 1980s to Early 1990s

The governmental White Paper on environment and development (Ministry of the Environment, 1989) sets out the following *domestic goals*, aimed at combining sustainable use and conservation of genetic resources, species and ecosystems (p.64):

1. To ensure ecologically sound exploitation of resources so that natural productivity and species diversity can be preserved for future generations; and, particularly, to conserve the habitats of endangered animal and plant species.[1]
2. To incorporate considerations for the environment and outdoor life into all relevant planning of building projects and utilisation of natural resources.
3. To safeguard a representative cross-section of Norwegian nature in the form of national parks, nature reserves, and so forth.
4. To ensure viable populations of salmon- and other domestic fish species, safeguarding these from over-exploitation, disease, pollution, and genetic contamination. 'Fish-farm-free zones' along the coast to protect wild populations of salmon are promised.

For the forest sector, the government spells out an additional main goal (ibid., p.122):

5. To secure the productivity of forest areas and protect large, continuous areas of importance to animal- and plant life through regulation and ecologically balanced forest management. An important policy measure for securing these goals is the Protection Plan for Coniferous Forests (*barskogplanen*), which was already initiated in 1981 (Ministry of the Environment, 1981).

Some of the most concrete policy measures are the Protection Plan for Coniferous Forests and the promise of 'fish-farm-free zones'. Moreover, in the years following the advent of the Convention on Biological Diversity, the Ministry of Agriculture set conserving biodiversity as a main goal in forest management, on equal footing with economic profit, in 1998-99 (Ministry of Agriculture, 1999).

In addition to the *domestic goals*, White Paper 46 (Ministry of the Environment, 1989, p.113) envisages several biodiversity-related goals aimed at external activities – that is, *international goals*:

1. To support the work to establish a global convention for the conservation of biological diversity.
2. To support international activities that give developing countries a more equitable share of the benefits accruing from the use of genetic resources.
3. To support development assistance organisations in increasing their attention and support for conservation and management of species and ecosystems.

The Norwegian goals build on the view that conservation and sustainable use should replace the traditional 'preservation' approach to the management of species and habitats (inherent in earlier international agreements). The new ideas imply an increased acceptance of human activities as being compatible with and necessary to nature and resources management, and are in line with the concept of sustainable development in the Brundtland Report (World Commission on Environment and Development, 1987). Compared to, for instance, emission targets associated with pollution control measures, the goals for nature management are in general easier in terms of attracting political support but more difficult in terms of measuring successful goal attainment.

Goal Attainments: 10 Years After Rio

Basically, it would seem that there are hardly any insurmountable hurdles against achieving the Norwegian *domestic goals* for biodiversity. Norway is a rich and sparsely populated country with relatively large areas remaining outside intensive land use. Adding will to ability, Norway has been at the forefront in international biodiversity negotiations. Accordingly, a number of achievements have indeed been made not least with regard to developing surveillance strategies and mapping biological diversity (Ministry of the Environment, 2001, p.128). The new management system for biological diversity involves identifying areas of great value for biodiversity and establishing a species database. Efforts have also been made to extend scientific knowledge. The Research Council of Norway has established two subsequent research programmes on biological diversity and, in cooperation with UNEP, Norway has hosted five international conferences (the Trondheim Conferences) linked to the follow-up of the CBD.

Goal attainment, however, does not increase in correlation with the accumulation of knowledge about biodiversity. On the legal dimension, a number of legal acts have received a face-lift, but their practical implications seem to be relatively small. The Government has appointed an expert group assigned to examine Norwegian biodiversity legislation and strengthen legal measures for the protection of biodiversity in Norway. The group was established in 2001 and must report back by spring 2004. Broadly speaking, however, the lack of impact from legal adjustments seems to be tied to an accompanying lack of economic incentives. Norwegian goal attainment in this policy field is generally connected to the meagre financial resources provided for each policy measure. In spite of Norway's benign starting point, the main threats to Norwegian biodiversity mirror those in most other parts of the world, with habitat fragmentation and deterioration and introduction of alien species high on the list.

With regard to *habitat conservation* in general (domestic goals 1 and 3), the implementation score is medium to high depending on which aspect of biodiversity is under scrutiny. Norway still boasts relatively large areas of wildlife in terms of established national parks, and the protected land area has recently been increased.[2] Protected areas currently add up to about ten per cent of Norway's land area, excluding Spitsbergen (Ministry of the Environment, 2002). In the OECD as a whole, the total amount of land protected is about ten per cent, which is equal to scientific recommendations (OECD, 2001a). If all plans for protected areas go through, total protected Norwegian land area will amount to 12-13 per cent. Still, the areas with the largest diversity of plants and animals are poorly represented in these plans, as mountain areas dominate (Ministry of the Environment, 2001, p.135; Ministry of the Environment, 2002). Moreover, the remaining wilderness areas have shrunken dramatically, from 48 per cent in 1900, 34 per cent in 1940, to 12 per cent in 1995 (Ministry of the Environment, 1996). Endangered species in the Norwegian flora and fauna still suffer from inadequate habitat protection. This is clearly contrary to Norway's domestic goals as well as its international goals.

Following on the same note for *forest management and protection* (domestic goals 1, 3, and 5), the verdict would seem to be a low score on goal attainment. There has been massive critique of the Norwegian failure to conserve the scientifically recommended amount of coniferous forests through the Protection Plan for Coniferous Forests. The scientific recommendations from the Norwegian Institute of Nature Research (Norsk institutt for naturforskning - NINA), among many others, hold that five per cent of the productive forest must be conserved. So far, only one per cent has received protection. This is contrary to the Norwegian explicit goal of adhering to scientific advice about resource management (see Chapter 1) and contrary to the domestic, concrete goals for nature- and forest management. The Norwegian budget for protection of coniferous forests has been approximately NOK 35 million a year,[3] while Sweden and Finland reportedly spend an annual NOK 500 million and NOK 460 million, respectively.[4] During the same timeframe, 'perverse' subsidies for the building of forest roads have amounted to around NOK 100 million yearly (see Appendix 1).

In terms of *species management* (domestic goals 1, 2, and 4), goal attainment ranges from medium to low. A major touchstone of the quality of Norwegian biodiversity policies will probably be related to how we manage the wild salmon. Here, the domestic goals were explicit (domestic goal 4). The Norwegian wild salmon is endemic and increasingly faces such threats as salmon louse from fish farms and escapees from the bred stocks.[5] Repeated studies show that the effect of escaped farm salmon on wild salmon is negative.[6] In 1999, the *Wild Salmon Committee* maintained that there is a need for *total protection* of nine salmon fjords and 50 salmon rivers (Ministry of the Environment, 1999). In response, the Labour Government (2000-2001) proposed a combination of *restrictions and conservation* aimed at fish farms in 22 fjords and 39 rivers. In 2002, while the protection plan for wild salmon was still pending final decision in the Norwegian Parliament (Storting), the Government offered 40 concessions for fish farms. In February 2003, the Parliament decided to give 13 national salmon fjords total protection as fish-farm-free zones – however, already established farms have been given an extension until 2011 to relocate.[7] Looking back at the 1988-89 objective of protecting the wild salmon by establishing fish-farm-free zones, it is evident that this 'vintage' goal has neither been updated nor achieved.

An example of more ambiguous goal attainment is found in the Norwegian management of cross-border populations of wolves (domestic goals 1 and 2). Only a few, tiny populations receive some kind of protection in Norway, and this has raised questions regarding the applicability of the precautionary principle and as to whether Norwegian management is in line with the Bern convention.[8] Co-manager Sweden, with considerably larger populations on their side of the border, has asked that the wolf be brought up as a *case of information* on the agenda of the meetings of the Bern Convention. This action involves a fair bit of political pressure on Norway and indicates that Norwegian policy for the protection of large predators has not been successful (Ulfstein, 2001).

In general, there is a growing realisation that the Norwegian legal system is still plagued by significant shortcomings. These shortcomings include a lack of legal protection of vulnerable habitats and species, regulation of introduction of alien species, regulation of access to and use of naturally occurring genetic resources, as well as liability rules for damage to biodiversity.[9] This realisation may be seen in relation to the growing acceptance and development of *the ecosystem approach*, with its focus on multiple use and local participation in decision-making processes, as well as identification and surveillance. Local participation may, however, be a double-edged sword. A great deal of ecological problems occur at the local level, where the Planning- and Building Act frequently illuminates the uneven struggle between biodiversity and economic concerns. This is particularly apparent with a view to protecting endangered species such as the big salamander (domestic goals 1 and 2). Norway is obliged under the Bern Convention to protect the big salamander and its habitat, but generally fails to do so. The prevailing explanation from officials at the municipality level is that there is no legal authority in the Wildlife Act to safeguard specific, single habitats of specific populations of endangered species.[10] The Planning- and Building Act provides such authority, but this is not followed by economic incentives. Following the work of the expert group on biodiversity legislation, a Biodiversity Act may remedy this situation.

In the international negotiations, Norway displayed high profiled, entrepreneurial leadership in the efforts to combine traditional nature preservation with sustainable use. Essentially, this means a greater focus on sectors such as agriculture, forestry, roads and transport, various trades and industries, and development assistance, and their impact on nature management (domestic goal 2). There is still a long way to go until this kind of sector integration is achieved in Norway, but the effect so far is an enhanced role of the Ministry of the Environment vis-à-vis other ministries.[11]

Against this backdrop, it can be concluded that Norway's attainment of *domestic goals* for biodiversity is medium to low. With regard to influencing *international regime outputs*, however, there is quite a high level of goal attainment. As the next section will show, Norway played a decisive role in the negotiation of the CBD and has similarly been active in international work to reduce the risk of alien organisms. Along with other Nordic countries, Norway pushed successfully for the idea that all countries share responsibility for biodiversity conservation and the objective of benefit sharing from use of genetic resources. The domestic *implementation* of these outputs shows, however, a different story. First and most notably, the CBD objective of benefit sharing from use of genetic resources has not yet received much attention in the Norwegian domestic implementation process. In contrast, Belgium and Denmark have made specific amendments in their domestic patent legislation in order to bring them more in line with the CBD objectives. Second, there has been little fine-tuning of goals for development assistance to support nature management strategies in line with the multilateral environmental agreements. While examples of specific assistance to support the establishment of environmental strategies in developing countries may be found,[12] the regular development assistance programmes in Norway, as in many other donor countries, are hardly in tune with the multilateral environmental agreements. There is a general lack of linkages between the political and operative level regarding environment and development in this sector (Ministry of Foreign Affairs, 1995; Rosendal, 2000).

A common trait of 'all' goals in relation to nature management is that there are few explicit standards by which to measure the success or failure of their implementation. This problem permeates the discussion of goal attainment, and will be dealt with by discussing concrete policy measures, or lack of such, in terms of the more general objectives outlined above. The following discussion explores the reasons why Norway has had a higher degree of goal attainment at the international level than at the domestic level, in spite of the apparent need, will, and ability to pursue ambitious domestic biodiversity policies.

Explaining Norway's International Achievements

This section investigates how Norway achieved such a high score on international goals by first examining the degree to which there was agreement between the involved ministries on the goals, and how various domestic institutional qualities affected the level of agreement between ministries. It looks into how regime qualities enhanced Norway's ability to shape regime outcomes and then examines the relationship between the core regime and related international institutions. The section demonstrates how the initial harmony between ministries is being challenged as other resource management regimes adopt the objectives of the CBD and enter into the implementation phase.

The CBD was signed at the Rio Conference on Environment and Development in June 1992 and entered into force less than two years later, in December 1993. It has attracted the ratification of all but a very few states (Afghanistan, Iraq, Somalia, and the US). It was only after a heated North-South debate that the CBD reached its three-fold, package-deal objective: To ensure conservation of biological diversity, sustainable use of its components, and the fair and equitable sharing of benefits from the use of genetic resources.

For Norway, the CBD negotiations provided a perfect arena in which to hone its green and development-friendly image to perfection. The political suitability of this performance was lent added sincerity by the great dedication and entrepreneurial spirit of the Nordic and Norwegian bureaucrats participating in the negotiation process (Koester, 1997; Svensson, 1993; Rosendal, 2000). Along with the other Nordic countries, Norway played a central role in bringing about the final compromise of the CBD. It was the Nordic countries that insisted throughout the negotiations that all countries have a common responsibility to share the costs of conservation. The Nordic countries saw themselves as *bridge builders* between the North and the South by arguing for a system that would allow a returning flow of benefits in compensation for the use of genetic resources from the South (Schei, 1997; Rosendal, 2000). In this endeavour, the Nordic countries supported and were supported by the developing countries (G77 and China), which had a high level of negotiation power in the regime formation process (Rosendal, 2000, p.124-5). Moreover, the principles that the CBD should include all biological diversity (both wild and domesticated), and that it should pertain to sustainable use along with conservation, constituted central Norwegian and Nordic goals (Schei, 1997; Svensson, 1993). At least as far as Norway is concerned, these objectives go back to the period preceding the CBD negotiations. The objectives set out in White Paper no. 46 (Ministry of the Environment, 1989) and the accompanying *Blue Book* (Ministry of Foreign Affairs, 1988) of the Ministry of Foreign Affairs were put into practice during the CBD negotiation phase.

As a result of the intense negotiations, the CBD is the first international treaty to address all aspects of biological diversity – meaning the diversity within (genetic diversity) and between species as well as the diversity of the world's ecosystems. It is also the first to couple this with strong socio-economic rhetoric. The bulk of the world's terrestrial species are found in tropical areas, not least tropical forests (UNEP, 1995, p.749). At the same time, it is primarily the developed countries of the North that possess the (bio)technological and economical capacity to reap – through intellectual property rights – the ever larger benefits from the genetic variety employed in the agribusiness and pharmaceutical industries. This situation made the issues of property rights and access to genetic resources and the responsibility for costly biodiversity conservation very central on the agenda during the CBD negotiation phase. A compromise was found by introducing the concept of 'equitable sharing' of benefits accruing from use of the genetic resources and by restating the principle that countries have national sovereign rights over the genetic resources within their boundaries.[13]

Domestic Institutions: Initial Inter-ministerial Harmony

Supporting the 'external' CBD principle of *equitable sharing* and hence, supporting developing countries with regard to the biodiversity issue came at a relatively low cost to Norway compared to the other developed countries. Norway had, and still has, a comparatively small biotechnology sector.[14] Hence, there are relatively limited domestic economic interests tied to the questions of intellectual property rights to genetic resources in Norway. In contrast, when the US backed out from signing the CBD in Rio, this was largely due to strong domestic pressure and fear that the CBD formulations might harm their domestic biotechnology sector (Rosendal, 1994; Porter, 1993; Raustiala, 1997).[15]

Also with regard to the 'internal' CBD principles of *conservation and sustainable use* there seemed to be few costs involved for Norway. Deforestation had long since ceased to be a problem, and there were still vast tracks of wilderness left in the mountainous country. The Norwegian delegation supported all efforts to enhance conservation and sustainable use of biodiversity and again found their policy goals reflected in the CBD negotiation output. Just before the CBD was adopted, the Ministry of Fisheries raised a lone voice of objection, suddenly realising that the treaty might interfere with resource management under their jurisdiction.[16] During the CBD negotiations, the ministries of foreign affairs, the environment, and agriculture participated in the delegation – although the Ministry of Agriculture put little effort into formulating its positions, appearing not to consider the biodiversity negotiations very relevant or potentially harmful to their own interests at that stage. The ministries of the environment and foreign affairs were in complete agreement. Coinciding with and identical to the goals in White Paper 46 (Ministry of the Environment, 1989), the Ministry of Foreign Affairs had set down specific goals linked to Norwegian *international* activities (Ministry of Foreign Affairs, 1988), pledging to support the work to establish a global convention for the conservation of biological diversity; support international activities that give developing countries a more equitable share from the benefits accruing from the use of genetic resources, and work for international development assistance organisations to increase their support for conservation and management of species and ecosystems.

At the time of writing – as the Rio outputs and instruments are coming of age – this inter-ministerial harmony has begun to show signs of stress. On the external dimension, the tensions have become apparent within the international forest management debate – for example, in the controversial question of a forest convention (Rosendal, 2001a). In spite of opposition from the ministries of the environment and foreign affairs, the Ministry of Agriculture succeeded in procuring a mandate allowing them to endorse a forest convention. Delegation members from the ministries of the environment and foreign affairs felt uncomfortable with this position, as their ministries had represented the high-profiled Norwegian position during the negotiation of the CBD objectives.[17] The crux of the debate pertains to the CBD principles on forest conservation, protection of traditional forest-related knowledge, and the rights of indigenous people on one hand, and economic interests in increased timber extraction on the other. Economic interests can be assumed to override concerns for forest conservation in the rich and forest-rich countries – those arguing for a forest convention.[18] For most of the forest-rich developing countries, however, the need to retain the equity principles from the CBD seems to have overrun this concern, as they are generally not in favour of a forest convention. In effect, the Norwegian (i.e. ministries of the environment and foreign affairs) goal of retaining the ***bridge-builder*** role is at odds with the Ministry of Agriculture's call for a forest convention.

Nevertheless, the overall impression is that officials from the ministries of the environment and foreign affairs enjoyed a great deal of latitude during the CBD negotiations. There may be two main reasons for this. First, not only were other relevant ministries disinterested at the time, but the media and public attention were busy watching the climate change debate being played out at their common stage in Rio. Second, like most of the domestic goals formulated for nature conservation and management, the CBD lacks concrete standards for implementation. On the same note, the CBD is largely a framework convention, which inherently spells vagueness until it is furnished with more concrete protocols, such as the Cartagena Protocol on Biosafety. This feature may certainly have made the CBD easier to accept by the negotiating parties, as well as disencumbering it from much sub-domestic resistance.

CBD – The Core International Institution: Facilitating Impact

Was Norway's ability to shape regime outcomes in any way hampered or helped by characteristics of the regime itself? Two features of the negotiation forum (provided by UNEP) may have helped to accommodate these goals. First, the negotiations culminated with the signing of the CBD at the UN Conference on Environment and Development (UNCED). The underlying norms of the UNCED setting were more clearly geared towards appeasing the developing countries than those of its predecessor, the UN Conference on the Human Environment (Stockholm, 1972). On the same note, timing seemed to play a role. The global public attention devoted to the UNCED may never have been greater, and this put strong pressure on top politicians to deliver. Second, the global participation in this negotiation forum has also been found to increase the scope for issue linking, including the necessity for the CBD to comprise both wild and domesticated material (Rosendal, 2000, p.130). Moreover, Koester (1997, p.179) points to the team spirit developed among the participants during the almost three years of regular meetings. In addition to its arena function, the UNEP Secretariat itself played a decisive role in the CBD negotiations. One after another, controversial items were integrated into the agenda and were supported by scientific input. Gradually, it became apparent to the negotiating parties that the links to intellectual property rights as well as the all-encompassing scope of the convention were necessary elements in order to get tropical, mega-diversity countries on board. The Executive Directive of UNEP, Dr Tolba, set up small working groups which focused on and boosted consensus on the remaining issues (Svensson, 1993). Consensus remains the main decision-making procedure for matters of substance. In addition, a number of compliance mechanisms were established with the potential to draw implementation in the same direction as the Nordic and the tropical countries' goals for biodiversity. Most significantly, the CBD was equipped with a monitoring mechanism in the form of national reporting and an incentive mechanism in the form of the Global Environmental Facility (GEF).

Briefly, we may ask how it was possible for this alliance between the Nordic and developing countries to win through in spite of persistent opposition from the US. One interpretation points to the vague and largely non-binding character of the CBD objectives, but this does not really explain the US refusal to ratify it. More important, the US was very much alone in its refusal to ratify the CBD. All the other OECD countries, including the EU, were quick to give their consent. Another interpretation is that the US stopped worrying about the output of the biodiversity negotiations at the time, as they were confident that their interests would be secured by the World Trade Organisation (WTO) and its Agreement on Trade-Related Aspects of Intellectual Property Rights (TRIPS) (Rosendal, 2000). Similarly, but seen from the 'other side' in the eyes of the US negotiation team, it was the developing countries that were trying to 'hollow out' the TRIPS Agreement by using the biodiversity convention (Raustiala, 1997, p.47). The following section looks further into such institutional interactions.

Other International Institutions: Regime Linkages

In the CBD negotiations, Norwegian diplomats followed up their policy goals from previous, related rounds in the UN Food and Agriculture Organization (FAO) – the negotiations on an International Undertaking for plant genetic resources in 1983 (renegotiated 1989, and renegotiated as a Treaty 2001). Other related regimes include the regional Bern Convention of 1979 (protection of wild species of plants and animals and their habitats), the Ramsar Convention on Wetlands (1971), the Bonn Convention of 1979 (on migrating species of animals), and the Convention on International Trade in Endangered Species (CITES, 1973). The regimes within the nature conservation cluster are legion, and generally draw in the same direction as that of the CBD (Rosendal, 2001c; Stokke and Thommessen, 2002). Mostly, they tend to specify action focused at particular areas or species. The CBD, in contrast, draws up more general objectives as well as more specific social principles for benefit-sharing, traditional knowledge and the rights of indigenous people.

At the time of negotiating the CBD, one potential stumbling block stood out at the international arena: the coordination of domestic positions vis-à-vis the Agreement on Trade-Related Aspects of Intellectual Property Rights (TRIPS Agreement) under the General Agreement on Tariffs and Trade (GATT) Uruguay Round (1988-94). The main objective of the TRIPS Agreement is to strengthen and expand the scope of intellectual property rights protection in all technological fields. This includes biotechnology, which is based on the utilisation of genetic resources. While concerns for equitable sharing and conservation constitute the core norms and principles engendered by the CBD, the TRIPS Agreement promotes the privatisation of genetic resources through intellectual property rights. The discussions on intellectual property rights in the Uruguay Round had significant impact on the question of access and benefit sharing pertaining to genetic resources, as debated within the framework of the CBD. While negotiating the CBD, governments were also engaged in the GATT Uruguay Round (1988-94), and this involved a fair bit of intra-ministerial discord. Norway's position in the TRIPS Agreement was in accord with that in the CBD negotiations; in both arenas Norway supported developing-country efforts to counter the increased strength and scope of patent protection within the biotechnological sector. The North-South divide arose to a large extent because patenting is a long and costly business that is primarily employed by large corporations.[19] At the time, warnings were issued that this position might lead to economic sanctions of Norway.[20] Nevertheless, Norway persisted in following their goals for the CBD – a strategy that may partly be explained by the relatively small Norwegian biotechnology sector. Since then, several Norwegian governments have had to grapple with the uneasy relationship between the CBD and TRIPS/WTO and, later on, the controversial EU directive on patents in biotechnology. The EU patent directive and the TRIPS Agreement have both been criticised for their potential to undermine the CBD objectives (Johnston and Yamin, 1997; Rosendal, 2001b). The EU directive has been criticised for not including sufficient provisions for equitable sharing, such as mandatory disclosure of where and how genetic material has been collected. Under the Bondevik I

Government, Norway submitted a Statement to the Court of Justice of the European Community, supporting the Netherlands in their strong opposition to the EU directive, among others on the grounds that its provisions are incompatible with a number of provisions under the CBD.[21] At the time of writing, Norway is in the process of accepting the EU patent directive and this is subject to major criticism from non-governmental organisations (NGOs). This has been one of the most controversial transpositions to date, resulting in the first abstention of a prime minister to accept an act of government.[22]

In addition, and at later stages, inter-ministerial coordination has had to clear some hurdles with regard to the international forest debate. First, this pertains to the Intergovernmental Forum on Forests (IFF) leading up to the UN Forum on Forests (UNFF). As the bulk of terrestrial species are found within forest ecosystems, a main controversy in the IFF debate centred on whether or not to establish a legally binding instrument – a Forest Convention – to protect the forest ecosystems.[23] Opponents of a Forest Convention claim that forests are already covered by existing institutions, such as the CBD, and that there is little reason to establish yet another. Among proponents, the national concern for increased profit from timber extraction can be assumed to override global concerns for forest conservation, which provides a reason to establish a separate convention (Kremen et al., 2000). As described further in the next section, this interaction has recently created some discomfort in the relationship between the Ministry of Environment and the Ministry of Agriculture.

Second, the international forest debate links up to the interaction between the CBD and the Kyoto Protocol of the UN Framework Convention on Climate Change (UNFCCC) (see Chapter 7). Forests represent a significant link between biodiversity and climate change, as a bulwark against atmospheric carbon dioxide (CO_2) build-up and a repository of the genetic heritage of the earth's flora and fauna. The link between forests as a 'sink' or carbon reservoir and forests as a repository for biological diversity may, however, be less clear-cut as far as remedies are concerned. The quickest and cheapest way to bind carbon may be to plant uniform, fast-growing softwoods. The homogeneous habitat created by this method is, however, hardly compatible with the objective of fighting biodiversity loss. This interaction was not widely appreciated at the time of negotiating the CBD and the UNFCCC. This issue will also be discussed further in the next section.

Article 2 of the Convention for the Protection of the Marine Environment of the Northeast Atlantic (OSPAR) refers to the CBD objectives and marine resource management, stating that the parties must take 'the necessary measures to protect and conserve the ecosystems and the biological diversity of the maritime area, and to restore, where practicable, marine areas which have been adversely affected.' This is also in line with Norwegian domestic goals, both the general goal of species and habitat conservation and the specific goal referring to safeguarding fish species. Issues relating to marine species management are also appearing more frequently on the CITES agenda, as fisheries management faces increasing challenges worldwide. Both regional and international fisheries management regimes (such as those of the FAO Fisheries Commission and CCAMLR)[24] seem to be incapable of rectifying the situation of failing stocks. Hence, proposals have been forthcoming to list for instance the North Sea herring under CITES. In the case of CITES listings, however, the Norwegian delegation, which is influenced by fisheries interests, tends to insist that CITES leave the management of marine species well alone.[25]

In sum, the regimes within the biodiversity conservation cluster broadly draw in the same direction. Among the regimes involved with approaches to access and intellectual property rights to genetic resources, there is currently a delicate balance between concerns for enhancing innovation through intellectual property rights on one hand, and a number of ethical, environmental and distributional concerns on the other (Rosendal, 2001b). Within those regimes that have a bearing on forest management, there is still plenty of room for tapping the synergistic potential, both internationally and at the domestic level. There is, however, scant evidence of activities tapping into this potential. Finally, with a view to marine and fisheries management, there is a tendency for related regimes to take on the CBD objectives. At the same time, there is a tendency for Norwegian interests to become de-linked from those of the core regime – as its material interests in fisheries management come into focus.

Summing Up

This presentation has examined the strong match between Norway's goals at the international level and the goals of the core regime. Norway, along with the developing countries and the other Nordic countries, had a breakthrough for their common position in the CBD negotiations. Several regime features enhanced Norway's ability to play the green, bridge-builder role. Moreover, there was not much disagreement regarding goals among the various ministries during the negotiation phase of the core regime. The match between the core regime and other regimes is a bit more complex, ranging from a harmonious 'biodiversity conservation cluster' to less consistent means and objectives within the issue areas of forest management, marine management, and intellectual property rights to genetic resources. The international institutions within these other clusters ('access and intellectual property rights cluster', 'marine cluster', 'forest cluster'), present domestic institutions with what may become incompatible demands in the implementation phase. The following section examines goal attainment in the implementation phase and how the CBD and other regimes have been employed by various stakeholders in this process.

Explaining Norwegian Domestic Achievements in the Forest Sector

While Norwegian efforts to attain a relatively ambitious regime on biodiversity have been successful, domestic efforts to implement these goals face more difficult obstacles. Within the limits of this chapter, it is, however, impossible to go into detail for all aspects pertaining to biodiversity policies in Norway.[26] Here, examination is limited to an in-depth case study of forest conservation and management. The background for this choice is that while forests accommodate more than half of the endangered and rare species in Norway, biodiversity goal attainment has been low in the forest sector. This section looks for factors that can help explain the mismatch between goal adoption and goal attainment in this sector. It highlights target group resistance to the implementation of biodiversity goals and how the legal and economic instruments for goal attainment have been too weak. For the more general discussion, examples will be added also from the management of specific species and policies with regard to intellectual property rights and equitable sharing.

International Institutions: Lack of Concerted Action

The CBD sets out obligations for the Contracting Parties to develop national strategies, plans and programmes for conservation and sustainable use; integrate conservation and sustainable use of biological diversity into relevant sector plans and policies; and develop systems of protected areas (Art. 6). The parties are required to identify components of biological diversity important for conservation and sustainable use; identify activities which have or are likely to have significant adverse impacts on conservation and sustainable use of biodiversity; and adopt economically and socially sound measures to act as incentives for conservation and sustainable use. Most of these obligations are mitigated by the legal formulations, *as far as possible* and *as appropriate*.

Nevertheless, the international focus on biodiversity and forest management, which also gave rise to the CBD, has no doubt put extra pressure on the Norwegian forest sector. The Norwegian national reports to the CBD Secretariat, the Norwegian Plans for Action on biological diversity and the White Paper on biological diversity (Ministry of the Environment, 2001) all came as responses to the CBD. New legislation is in the pipeline with regard to the proposed Biodiversity Act, and this is an explicit response to the CBD. Likewise, a new Forest act is considered necessary to follow up obligations under the CBD. In what may be seen as an effect of the CBD, as well as the Rio forest principles, the Ministry of the Environment added another 120 km^2 to the area covered under the *Protection Plan for Coniferous Forests*, increasing the total protection of productive fir forest areas from 0.8 per cent to just above one per cent. The slight adjustments in the forest policies made by the Ministry of Agriculture are probably as much a result of the Pan-European Helsinki Process on Forests, as they are a direct effect of the CBD. Then again, the Pan-European Process has also been prodded by the international focus on forests, which arose during the preparations for, and aftermath of, the UNCED Conference. *Living Forests*, a private initiative from the forest sector, is primarily a response to market pressure that was sharpened by NGOs that were given new and powerful arguments by multilateral environmental agreements such as the CBD.

More weight seems thus to have been put on the forest issue in response to the CBD, but it is hard to distinguish this effect from what would have taken place anyway due to domestic priorities. In 1997, White Paper No. 58 envisaged a number of policy instruments aimed at attaining the objectives in the CBD. In the spirit of the Green Tax Commission, it proposed the elimination of subsidies that might have a negative effect on biodiversity, as well as a 'nature tax' moulded on the 'polluter pays principle'. Compared to the field of pollution control, however, nature management seems rather immune to such instruments. It generally lends itself to less specific and dynamic goal formulations, it is hard to translate into operative policies, and the development of conservation plans is a long and painstaking process. Moreover, with a view to protecting specific species and habitats, international institutions such as the CBD are unlikely to provide much direct support to weak domestic legislation. At the domestic level, the present legal framework offers only nominal protection to threatened wildlife, and it has frequently been shown to be quite unsuited for protecting specific habitats and species. Lack of accompanying financial compensation renders such legal instruments weak and ineffective.

Domestic goal attainment with regard to biodiversity may in the end also be affected by other international regimes. The relationship between the CBD and the UNFCCC, more specifically the Kyoto Protocol, may turn out to have long-term effects for forest management. The effect is illustrated by the difference between appreciating the standing stocks of forests and the quality of the forest. The first leads attention to the fact that there is no actual deforestation in quantitative terms, and hence, no CO_2 emissions. The second points to the change of tree species on the Norwegian western coast, which may have negative effects in terms of loss of biodiversity. The problem for biodiversity is that the CBD generally lacks concrete standards by which to measure implementation activities. It thus becomes vulnerable to economic incentives that may have adverse effects on biodiversity, such as those that may emanate from the Kyoto Protocol. In a report on forests and the potential to mitigate climate change from the Ministry of Agriculture (Aalde et al., 1997), a number of measures are envisaged. Most importantly, these include afforestation, reforestation and change of tree species, all of which represent potential threats to biodiversity (Solås, 1998). At the same time, biologists argue that the most cost-efficient manner by which to enhance CO_2 sinks is delayed harvesting (de Wit and Kvindesland, 1999; Solberg, 1997; Solås, 1998).[27] Biologists ask why the cost-efficient measures, which are also beneficial for biodiversity, have not been proposed by Norwegian authorities (Solås, 1998; Solås and Aanderaa, 1998). The same question is directed at the international debate on climate change, with the proposal of making forest conservation eligible to carbon credits along the line of afforestation and reforestation.[28] While Norway continues to pursue green goals in the CBD, this is much less the case within the UNFCCC. This may in turn have reduced Norway's scope for advocating the synergies between the two institutions. Part of the reluctance to link biodiversity conservation to the credit system of Kyoto stems from the Ministry of the Environment. The ministry's department on biodiversity and culture made repeated efforts to insert language to avoid conflict with the CBD objectives within the UNFCCC negotiation mandate. These efforts were, however, initially met with reluctance by the section in charge of those negotiations – the ministry's climate section.[29] Delegation members from the climate section were apprehensive about opening up for new interpretations of the delicate Kyoto deal and wanted to secure agreement on what was already there, before adding new elements.[30] It should be pointed out that the effect of the UNFCCC has mainly been to reduce Norwegian efforts to pursue biodiversity goals in other international fora; it has hardly had any noticeable effects on domestic policies. This situation is not unique for Norwegian environmental authorities – German biodiversity negotiators make the same 'complaints'.[31] While this situation characterised the UNFCCC meetings in The Hague (2000) and Bonn (2001), Norwegian reluctance was largely overcome at the seventh Conference of the Parties (COP7) in Marrakech (2001).[32] While forest conservation is a win-win situation for both climate and biodiversity, it is not favoured by the 'climate NGOs', who fear a watering down of the obligations of the Kyoto Protocol. Looking at the climate change process it is apparent that, so far, the scope for synergies remains high and largely untapped (Rosendal, 2001a). The long-term implications for domestic goal attainment may be that short-term management concerns will continue to override concerns for protecting specific species and habitats.

Finally, the effects of EU legislative instruments on domestic goal attainment must be examined. Basically, nature conservation, along with agriculture and fisheries, were excluded from the European Economic Area agreement (EEA). Norway wanted sovereign control over the management of natural resources. In effect, the most central biodiversity instrument of the EU – the Habitat Directive of 1992 – does not apply directly to Norwegian policies.[33] The basic Norwegian goal is that Norwegian politics in this area should equal that of the EU,[34] but this has not been achieved (Hustad, 2002; Reinvang, 2003). Essentially, this may be explained by a difference in emphasis on underlying principles in EU and Norwegian nature conservation. The principal difference can be illustrated by the CBD negotiations when several EU member countries preferred listings over habitats and threatened species while the Nordic countries advocated environmental integration and sustainable use also outside the 'islands' of protected areas. In practice, Norway may have come further in formulating goals for sector integration, but the EU is in the lead on legal and economic instruments that actually protect specific species and habitats. Several indicators imply that EU biodiversity policies to a greater extent than the Norwegian ones have bridged the gap between goal formulation and implementation. The *Natura 2000* plan of the Habitat Directive includes concrete demands for the protection of habitats of threatened species. Over 15 per cent of EU territory has been proposed for inclusion in this plan, which is estimated to cost between €3.4 billion and €5.7 billion yearly (between NOK 25 and 41 billion).[35] Moreover, the Habitat Directive has a more urgent deadline compared to Norwegian protection plans, as it aims for implementation by 2004. Funding is more than likely to be problematic, but in contrast to equivalent Norwegian legislation, the *Natura 2000* of the Habitat Directive has legal teeth. One example is a ruling by the European Court of Justice, where Greece was found guilty of and fined for being late in protecting the sea turtle habitat on Zakinthos Island.[36] Against this backdrop, it may be argued that realisation of policy goals for protected areas could have been improved if Norway had had to abide by the Habitat Directive.

There are two avenues through which EU biodiversity policies may still affect Norwegian policy-making. First, the *Natura 2000* of the Habitat Directive has had major impact on the development of the Bern Convention with its network of protected areas (the *Emerald Network*), and thus indirectly affects Norway. Second, Norway is in the process of endorsing the EU Water Framework Directive, which does constitute a part of the EEA agreement and which requires stricter standards and deadlines for the implementation of the Habitat Directive. This involves those protected sites under the *Natura 2000* network, which are 'dependent upon water'. This will necessitate a response from Norway, which is lagging behind relative to the *Natura 2000* and also in terms of the less stringent *Emerald Network* under the Bern Convention.[37] As it turns out, however, the current Norwegian response is to identify the habitat-related standards in the Water Directive. As these particular standards are not part of EEA, it will be possible for Norway to avoid compliance.

Domestic Institutions: Weak Policy Instruments, Fragmentation and Scientific Discord

The most important acts that can be used for area protection in Norway are the Nature Conservation Act and the Svalbard (Spitsbergen) Environmental Act. Conservation initiatives are also included in other acts, such as the Planning- and Building Act, the Wildlife Act, the Act Relating to Salmonids and Freshwater Fish, the Cultural Heritage Act, the Act on Saltwater Fishes, and the Aquaculture Act. Within this legal framework, the central authorities, the counties, and the municipalities have different responsibilities when it comes to land-use planning.

The body primarily responsible for biodiversity is the Ministry of the Environment, which has the overall responsibility for habitat conservation, including the *Protection Plan for Coniferous Forests*. It also has established the goal that, by 2003, all municipalities in Norway shall map and evaluate important areas for biodiversity, based on a handbook drawn up by the Directorate for Nature Management (Ministry of the Environment, 1996). At the same time, it is the Ministry of Agriculture that is basically responsible for forest management – including establishment of legal and economic instruments in the forest sector and for the scientific approach through its newly developed *Environmental Registrations in Forestry – Biological Diversity*. In addition to activities under the auspices of these ministries, activities introduced by the commercial interests of the forest sector – particularly the project *Living Forests* – are also relevant.

Forest management is partly a matter of protecting selected parts of the forested areas and partly a matter of determining how biodiversity is treated within the management of productive forests. Hence, forest management necessarily implies facing the contention between preservation and sustainable use, and Norwegian forest policies have been subject to criticism on both fronts. The challenges primarily ensue from the division of responsibility between sector ministries and environmental authorities, and often involve discord about the interpretation of scientific recommendations.

Let us first look into the role of science. As an example of scientific discord, the OECD (1993; 2001b) has repeatedly criticised the Norwegian *Protection Plan for Coniferous Forests*, pointing out that one per cent is well below the minimum level of preservation recommended by science. In response, the Ministry of

Agriculture has questioned both the validity and the utility of such scientific recommendations.[38] In comparison, Sweden is in the process of protecting about five per cent of their productive forest area, and their long-term goals are linked to the scientific recommendations that between nine and 16 per cent must be protected. The scientific advice from the World Conservation Union (IUCN) is to protect at least ten per cent while the Nordic Council of Ministers recommends protection of 5-15 per cent of the productive forest. In the case of forests, the resort to scientific uncertainty as an argument for delay stems from the central administration itself – the Ministry of Agriculture. A similar trend can be observed within the fish-farming sector. Here it is primarily the trade interests that raise questions about the scientific validity of conservation plans,[39] but also within this field there seem to be close ties between the trade and sector ministry.[40]

Scientific discord may spill over into discord about budgetary decisions. A central critique of the *economic* instruments employed in the forest sector has been that they give rise to 'perverse subsidies' (Gulbrandsen, 2001). In 2000 for instance, NOK 128 million was allocated from the government in order to encourage road construction through forested areas, changing tree species, and supporting timber harvesting in remote and difficult areas.[41] Funding for such developments in the forest sector generally exceed the total amount of funding for protected areas, including coniferous forests. As late as in the White Paper of 1998/99 on forests, the Ministry of Agriculture envisages a 20-30 per cent increase in forest roads. The Norwegian Institute for Forest Research concludes that of the 1,619 forest-dwelling species on the Norwegian Red List, 17 per cent are directly threatened by forestry, while only 13 per cent are clearly at risk due to factors other than forestry (Ministry of the Environment, 2001, p.118). According to the Ministry of the Environment, the rapid reduction in wilderness areas is largely a consequence of forest roads built by the forest sector (ibid., p.37; Ministry of the Environment, 1996, p.33). Commenting on these findings, the Ministry of Agriculture again questioned the scientific validity and usefulness of the Red Lists (Ministry of the Environment, 2001, p.118), which may partly explain why it recently introduced its own system: *Environmental Registrations in Forestry – Biodiversity* (Gulbrandsen, 2001 and 2003). Within this system, the responsibility for identifying and paying for protection of key biotopes in productive forests is left to the forestry planners, which basically means increased control for the forest owners themselves (Gulbrandsen, 2001 and 2003).[42] The Ministry of Agriculture's *Environmental Registrations in Forestry* also differs from the national surveillance strategies and biodiversity mapping that has already been developed (Ministry of the Environment, 2001), first by its lack of transparency and second by its much smaller scope in terms of biotopes and criteria. In contrast, Sweden has employed independent biologists to carry out this registration, which is now mostly complete. In Norway, no results are available at this time.

The fragmented responsibility, and in turn influence, between ministries can be identified also through more general budgetary terms. The low score on forest protection must be seen in relation to the sparse financial allocations – roughly one tenth of the Swedish and Finnish budgets. In the White Paper on biodiversity (ibid., p.137), the Ministry of the Environment estimated that the present *Protection Plan for Coniferous Forests* (which stipulates protection of one per cent of forest area) would involve an added cost of about NOK 680 million. Norwegian environmental NGOs estimate that the scientifically recommended goal of protecting *five* per cent would cost between NOK four and six billion.[43] Likewise, NINA and Skogforsk (*Norsk institutt for skogforskning*: Norwegian Institute for Forest Research) estimate that the costs of increasing protection to 4.5 per cent would be NOK 3.5 billion (NINA and Skogforsk, 2002).[44] So far, no government budgets have included funding for such an expansion of the protection plan. These figures indicate that representative habitat conservation in Norway suffers badly from lack of funding.

Finally, the contested division of responsibilities between ministries is also reflected in the legal instruments employed by the Ministry of Agriculture in the forest sector. Basically, the *legal* framework provides the forest sector itself with the principle of 'freedom with responsibility' (Ministry of Agriculture, 1989, p.33). Principles concerning biodiversity have recently been incorporated in the forest legislation (Ministry of Agriculture, 1999). These principles remain, however, vague and ineffectual, when confronted with concerns for employment, district policy and economic growth. Similar fragmentation can also be found in other sectors, such as fish farming, where as many as three ministries are engaged in various parts of management. In the case of salmon, the Ministry of the Environment is responsible for managing the wild species; the Ministry of Fisheries deals with farmed fish; and the Ministry of Agriculture becomes engaged in case of diseases in both wild and bred stocks. These examples indicate that the call for unified, sector-integrated action still has a long way to go.

Ministries, the forest sector and the NGO community all agree with regard to which forest activity has been most important for changing behaviour in a more environmentally friendly direction (Gulbrandsen, 2001 and 2003). In response to strong pressure from NGOs and consumers, this activity was introduced by the economic actors in the forest sector itself in 1995 under the heading *Living Forests*. The aim of *Living Forests* was to increase credibility among domestic and foreign consumers that Norwegian forest products were deriving from sustainably managed forests. In 1997, *Living Forests* introduced certification based on standards and guidelines for sustainable forest management, including the registration of key biotopes. The project initially included NGOs such as the WWF, SABIMA,[45] and the Norwegian Society for the Conservation of Nature (*Naturvernforbundet*). These NGOs later resigned from the cooperation on account of the many infringements of what they regard as already weak standards, especially with regard to key biotopes (SABIMA, 2003). The aspect of key biotope registrations is further weakened in the new method of *Environmental Registrations in Forestry*. The NGOs were also disappointed that *Living Forests* chose to certify their products through the self-imposed ISO 14001, and later on the Pan European Forest Certificate (PEFC) standards rather than those of the independent Forest Stewardship Council (FSC). Once more, this differs greatly from the Swedish situation, where most of the large forest owners use FSC certificates, and the rest are in the process of harmonising 'upwards' to FSC standards (Gulbrandsen, 2003).

Foreign and Domestic Efforts: Private Sector Main Driving Force for Environmental Change

The main questions to be addressed in this section concern how target groups became involved in goal formulation and the process of goal attainment. The time dimension is also considered by examining whether new regimes may have had an impact on goal attainment.

Norway clearly receives a much higher score on international goal attainment compared to domestic achievements, in spite of the two starting out as such compatible companions. This coincides with the observation that there was a high degree of domestic consensus during the international regime formation phase, followed by much less coherent domestic positions in the implementation phase. For a large part, this can be traced back to 'vertical disintegration' explanations. Several sectors had scarce idea that they would end up being affected by implementation activities. Thus they were brought into the policy process at too late a stage and greatly lack commitment to follow up domestic obligations. Section three argued that the Ministry of Agriculture did not appear to consider the biodiversity negotiations as relevant or potentially harmful to its own interests during the CBD negotiations. A few years later, this situation was changing as the forestry issue reappeared at the international stage with the IPF/UNFF process.[46] The Ministry of Agriculture has been very concerned about playing the leading role in these rounds – rather than leaving it to the ministries of foreign affairs and the environment – and they have largely succeeded in this endeavour. The change of view within the Ministry of Agriculture with regard to biodiversity coincided with an intra-ministerial change in representation. During the CBD negotiation phase, the Ministry of Agriculture was represented by officials from its agricultural department (*Landbrukspolitisk avdeling*), and at that time there were no signs of discord between the ministries of agriculture and the environment. This situation changed in the implementation phase when the Ministry of Agriculture was represented by its forest department (*skogavdelingen*), which disagrees with the Ministry of the Environment on a number of items.[47] The Ministry of Agriculture, or more precisely its forest department, now seems to regret that it did not participate more actively during the CBD negotiations. It points out how the forest sector was not adequately represented at the time and, while the Ministry of Agriculture is actively engaged in the CBD forest programme at present, it still lacks 'ownership' of this process. In comparison, the Ministry of Agriculture has a stronger sense of ownership for the United Nations Forum on Forests (UNFF), which is aimed purely at the development of forest policies. On the 'negative' side, from an environmental point of view, the UNFF to a greater extent than the CBD accommodates the interests of the forest sector, and it has been criticised for delaying the process of resolving the environmental problems in the forest sector.[48] Hence, forest management frequently becomes a battleground between the two ministries.[49] When the economic instruments employed by the Ministry of Agriculture are described as 'perverse subsidies' by sources from the Ministry of the Environment, this is basically because economic benefits still rank above environmental considerations in forest legislation.

If the forest sector and its target groups had been 'realistically' involved at an earlier stage, the output of the CBD negotiations, as well as the government's domestic goals, might well have looked very different. Another significant point relating to this part of the explanation involves financial priorities. When target groups at local levels resist making the preferred behaviour change, this may also be interpreted in light of the resistance at the central level to allocate the necessary economical incentives and compensation. The scarce public funding is essentially explained as an inherent part of 'freedom with responsibility', as pursued by the forest sector itself.[50] The major changes in the forest sector have come about as a consequence of private initiatives, such as *Living Forests*. The 2004 proposal for 'voluntary conservation', where the forest owners themselves offer up areas for protection, is not inherently different in this respect, as it also depends on sufficient public funding as well as improved biological representation in order to increase goal attainment.

In addition, there are other global regimes to consider. It seems a fair verdict that the climate change issue has received much more attention in Norway compared to biodiversity, also after the public focus on environmental concerns in general has waned. Moreover, the UNFCCC may possibly, but not necessarily, have the potential to work against the biodiversity objectives for forest management.[51] The forest sector may, or may not, choose to utilise this potential to increase financial support for changing tree species, afforestation, and reforestation activities. Similar discrepancies in environmental goals can be envisaged between the need for clean energy (hydropower) and habitat protection (Eikeland, forthcoming). At the same time, the synergistic international regimes within the nature conservation cluster have not been much more effective than the CBD in enhancing Norwegian goal attainment. A possible exception to this picture may be the EU Habitat Directive, which has some parallels to the Norwegian plans for environmental registrations. While nature conservation and management issues are not part of the EEA agreement, EU policies may still have an effect on relevant policies in Norway – if nothing else, for competitive reasons. So far and in general, there has not been much direct impact from other international institutions on Norwegian goal attainment.

Other Factors Confounding Difficulties: Scant Technological Solutions and Protection of Rural Communities

In addition to institutional factors at the international and national level, several other factors may influence the implementation of environmental goals and obligations. This section briefly discusses the potential impacts on biodiversity policy from characteristics of the *problem type*, including the general *interest structure*, at the domestic level.

The loss of biodiversity is different from issues of pollution control on a number of aspects relating to *problem type*. First, while pollution issues have often been aided by external shocks to draw attention to their importance, such as algae invasions or ozone holes, similar events are scarce in the biodiversity issue.

Moreover, if we can talk about external shocks involved in the biodiversity issue, they have occurred far from Norway, such as tropical deforestation in the Amazon rainforest. In 2000, shooting wolves from helicopters did bring some flavour into the Norwegian biodiversity debate, but this did not have a major mobilising effect. Second, more than any other environmental issue area, biodiversity is characterised by highly concentrated costs and highly distributed benefits. The same is true for the climate change issue, but goal attainment in climate change is facilitated by more clear-cut standards. Moreover, the target groups of climate change and other pollution issues partly consist of companies, which are easier to control than the large number of individual farmers and foresters. The Norwegian forest sector is comprised of a comparatively large number (80 per cent) of small-scale, private forest owners. Even if sufficient money had been provided for conservation and sustainable use, this group would hardly have been easily manageable. Third, pollution control may be facilitated by innovative environmental technology. It is less apparent how technology may reverse or restrain changes in land use and the fragmentation of habitats. Hence, biodiversity emerges as an exceptionally hard area in which to change the behaviour of target groups.

Turning to interest structure, there are clear indications of the traditional lines of *political cleavage between conservation and growth*. The controversies in the forest sector, as well as the fish farming industry, largely follow this cleavage, and it may also have explanatory force in the EU patent question. Some of the cases are further complicated by an additional, coinciding cleavage with equally long standing traditions in Norwegian politics; the *centre-periphery cleavage*. This line of cleavage adds to the tensions in the forest sector as well as livestock farming, from traditional sheep farms to the more novel salmon farming industries. The greater obstacle from the point of view of wolves and whales (see Chapter 3 on whaling) is probably the political clout of rather small-scale farmers and fishers. The political influence of these groups derives not from numbers or economic power, but from the symbolic strength of Norwegian rural community policy. Compared to the rural interests involved in the wolf case, the economic power behind the fish-farm industry and the forestry sector is much higher, but these sectors may be more vulnerable and receptive to the greening of consumer power and market pressure. In this perspective, the threats to wild salmon and forest biodiversity may be easier to mitigate than the threats to the wolf.

Compounding these difficulties is the deployment of environmental experts at local levels of government. This involved dismissing entirely or reducing the responsibilities of environmental consultants in 18 of Norway's 19 counties and in 251 of the 435 municipalities since 1997 (Bjørnes and Lafferty, 2000). The great reductions were a result of the decision to stop the 'targeting' of state money to environmental consultants at the municipal level. The effect has been to rid the local levels of a great deal of biological competence as well as capacity to pursue environmental goals (ibid.).

Any analysis of environmental policy must also take into account the fact that environmental concern has sunk drastically since the early 1990s and the Rio Conference. The most recent developments in Norwegian environmental policies in general can be pinpointed to a single individual – Environmental Minister Børge

Brende. His role has been important for setting the environmental wheels in motion, although arguably more so in terms of increasing the speed of classical wildlife protection than with a view to the more costly integration of environmental concerns in the various sectors. Nevertheless, this is a reminder of the significance of individuals, which is often overlooked in institutional policy analysis.

Norwegian Biodiversity Policy – Main Lessons

Let us sum up how the main domestic and international factors have affected goal attainment. The limited goal attainment of Norwegian domestic biodiversity policy is largely accounted for by *domestic institutional factors*. Basically, three aspects characterize the policy instruments of the forest sector. First, both the legal and the economic instruments applied for habitat protection are insufficient to achieve the stated goals. Second, the economic instruments applied within the forest sector do little to constrain the behaviour of target groups; indeed, they may even lead to increased loss of biodiversity. Third, there are domestic controversies related to the scientific basis for environmental recommendations as well as the scientific soundness of the instruments applied (*Environmental Registrations in Forestry* and *Living Forests*). In addition, biodiversity policies generally lack the pull from technological innovations and are also confounded by the Norwegian policy of protecting rural communities, which may conflict with the protection of biodiversity. Adding this to the general lack of economic and legal factors pulling for biodiversity conservation in Norway, the relatively low goal attainment in the area hardly needs additional explanation.

More specifically, this chapter has shown how vertical and horizontal disintegration explains much of the implementation problems encountered within the biodiversity issue area. Moving *vertically* from the international negotiations and agreements to the domestic implementation phase, it is hardly surprising that new stumbling blocks appear. Local levels of administration are left with the difficult role of arbiter between costly environmental concerns and ready economic gains. One of the most significant barriers to establishing a biodiversity policy in line with both domestic and international goals springs from the environmental weakness of authorities at the local levels. Bureaucrats at local levels of administration tend to complain that the central ministries present them with incompatible demands and those emanating from the Ministry of the Environment often lose out in this struggle.[52] It is generally easier to find allies to further industrial development than to support environmental concerns. *Horizontally*, sector ministries come to realise that their interests may be compromised as central obligations are translated into practical politics. The corporate interests that inhabit this sector are yet another step removed from the multilateral environmental agreements. The Norwegian forest sector, from large state-owned to small-scale privately owned enterprises, has received meagre incentives, other than market pressure, to follow up the goals of the CBD and the domestic goals that support it. The ministries of the environment and agriculture both agree that lack of public funding is partly a consequence of the self-chosen strategy of 'freedom with responsibility' in the private forest sector.[53] Stubbornness in the private sector would not, however, seem to absolve the political authorities from an overall responsibility for attaining domestic biodiversity goals.

Norwegian domestic policies on forest biodiversity have not been subject to much international crossfire, but neither has there been much environmental pressure by the relatively weak international institutions. Granted, the Ministry of the Environment would probably have achieved less in the absence of the CBD, but it still has a solitary job in trying to protect forest species and habitats. This situation might have been improved upon if Norway had commitments under the biodiversity policy framework developed under the EU. At the same time, Norway continues to pursue biodiversity-related processes at the international level, and this has illuminated problems of coordination both within and between sector ministries. These problems can be expected to increase as issues relating to forestry- and fisheries management are more frequently being dealt with in international fora.

Seen in light of the five general Norwegian environmental goals (Chapter 1), the biodiversity policies may be *cost-efficient* but they hardly accomplish the aim of *leading by example*. Another goal concerns *respect for international conservation rules* and adherence to international obligations. Norwegian behaviour in the case of whaling is hardly in the spirit of the CBD (see Chapter 3), but apart from this, Norwegian inability to comply is found in action rather than words. Furthermore, Norwegian biodiversity policies generally fall short of the objective of *adhering to scientific advice*. Instead, a practice of questioning and undermining the credibility of scientific advice is developing within several sectors, most notably those of forestry and fish farming. In both cases, the notion of the precautionary principle hardly enters the debate. While there is widespread scientific agreement about the great threats to biodiversity and accumulating biological knowledge, this is not accompanied by corresponding implementation of the policy goals for reducing the loss of biodiversity. The same is true with regard to achieving the aim of *promoting development in the Third World*. Norway has done a considerable job in promoting developing country interests at the international level, but still lacks a domestic, legal framework for access and benefit-sharing pertaining to use of genetic resources in line with the CBD objectives. International crossfire in the biodiversity issue area is hardly intensive enough to excuse the relatively low goal attainment. The main explanations for these shortcomings are found at the domestic level. Repeatedly, Norwegian negotiators signal will and ability to make great environmental leaps at the international stage, only to be stopped short by insufficient domestic funding. It is hard to believe in flying starts for environmental policies when the impressive international bungee jumps are persistently followed by financial belly flops at home.

Notes

1 One policy measure envisaged for securing these goals was the ensuing report on large predators (Ministry of the Environment, 1991).

2 By 2001, protected areas consisted of 18 National Parks, 97 Landscape Protected Areas (up from 76 in 1997), and 1441 Nature Reserves (up from 1172 in 1997) in Norway. The increase is rather high following the 1988-89 report, and it could thus be considered as high in terms of implementation score (Ministry of the Environment, 2000).

3 National budgets (St.t.prop. no. 1.) See appendix 1 on government subsidies to nature conservation.

4 www.noa.no/Saker/Skogbruk/Vern.htm.

5 During the last decade, the percentage of escapees has remained stable, ranging from 28 to 40 per cent (Fiske et al., 2001).

6 In 49 out of 667 salmon rivers, the populations of wild salmon have become extinct. In a further 76 rivers, the remaining populations are threatened by acidification and *Gyrodactylus Salaris*, and another third of all salmon rivers are affected by dam and electrification projects. (Hindar et al., 1991, and references therein; Fleming et al., 2000, and references therein; Hindar, 2001).

7 An additional 21 fjords and 37 rivers received a combination of *restrictions and conservation.* St.prp. no. 79 (2001–2002) and Innst.S.no. 134 (2002–2003).

8 A report by the OECD maintains that the most effective measure would involve setting aside sufficiently large areas in which livestock is banned and the protection of the wolf strictly controlled (OECD, 1993, p.40). As shown in Appendix 1, however, the bulk of government funding has been spent on damage costs in terms of compensation to livestock farmers. See also Ulfstein, 2001.

9 These shortcomings were identified in Ministry of the Environment, 2001, p.27.

10 www.noa.no/saker/bymiljø/brev.htm

11 Personal communication, Gunn Paulsen, the Directorate for Nature Management (DN), November 2001.

12 At the time of writing, there are ongoing projects in Costa Rica, Indonesia, Nepal, and South Africa. In addition, twelve projects in seven African countries have recently been finished (Eritrea, Ethiopia, Malawi, Tanzania, Uganda, Zambia, and Zimbabwe), http://environment.norad.no/centres.cfm.

13 Access to genetic resources requires *prior informed consent* and must be on *mutually agreed terms*. Thus, the CBD pits national sovereignty against the increasing use of patents by growing biotechnological sectors in the rich countries.

14 Ernst &Young (2001) provide a yearly ranking of the biotechnology sectors in Europe: According to the 2001 ranking, Norway shares the lowest ranking with Ireland and Spain, both in terms of production and in the number of enterprises. While Sweden has 150 biotechnology enterprises, Norway has about 30.

15 In 1994 the US signed the CBD, followed by an interpretative statement. To date, however, it has still not ratified the CBD.

16 Ministry of Fisheries Note, 19 February, 1992. Reassurances were, however, immediately forthcoming and the special call for removing marine living resources from the CBD did not at any point enter the international negotiation mandate. Personal communication from Norwegian delegation-member to the CBD negotiations, Jan P. Borring, MoE.

17 This section builds partly on interview with Anne Marie Skjold, the Ministry of Foreign Affairs (20 January 2000) and partly on the author's own observations during the IFF-4 meeting (New York, January 2000). See also Rosendal (2001a).

18 Most of the countries in favour of a forest convention are very forest-rich in the case of industrialised countries, and moderately to very forest-poor in the case of developing countries.

19 Transnational corporations hold 90 per cent of all technology and product patents (Gleckman, 1995). The developing world holds no more than one to three per cent of all patents worldwide (UNDP, 1999 and 2000).

20 Note from Jørgen Smith, Director of the Norwegian Patent Board as of 1992. Personal communication, Jan Borring, Ministry of Environment, December 2001.

21 Oslo, 15 March 1999; *Statement In Intervention By The Norwegian Government In Case C-377/98.*

22 Prime Minister Kjell Magne Bondevik of the Christian Democrats (the Bondevik II Government) maintained that he could not go against his conviction in the case.

23 Forests have been estimated to contain between 50 and 85 per cent of the world's terrestrial species (UNEP, 1995, p.749).

24 The Convention on the Conservation of Antarctic Marine Living Resources.

25 Author's personal observation at CITES, COP 12 meeting in Santiago, Chile 2002.

26 An overview of sector responsibilities and the coordination within and between the central ministries and the administrative levels can be found in *White Paper to Parliament No. 42* (Ministry of the Environment, 2000-01).

27 An increase in the rotation length, which may be achieved through the conservation of key-biotopes and forest edge zones, is also compatible with conservation of biodiversity (Solås and Aanderaa, 1998). The largest carbon stocks (50-60 per cent) are found in organic soil, and the planting or regeneration of mixed forest provides the highest carbon stocks in the forest floor (WRI, 1991). Conversely, measures that may negatively affect carbon stocks in soil are strongly mechanised site preparations, intense harvesting such as clear cutting, and draining and planting of peat lands (de Wit & Kvindesland, 1999).

28 While conservation of natural forests has very positive impacts on biodiversity, the inclusion of avoided deforestation in Article 3.4 and 12 of the Kyoto Protocol may give rise to methodological problems, such as questions of permanence, leakage, and baseline setting (FCCC/SBSTA/2001/MISC.3). Similar methodological problems are, however, part of most activities aimed at reducing greenhouse gasses.

29 *Seksjon for internasjonale forhandlinger om klima og sur nedbør* (Section for international negotiations on climate and acid rain).

30 Interview with Håvard Thoresen, director of the Ministry of the Environment's climate section, 21 June 2000.

31 Dr. Horst Korn, Federal Agency for Nature Conservation, Germany: 'Climate people told us frankly that the negotiations for the Kyoto Protocol were so important that they unfortunately did not have any time to consider biodiversity issues (for them a very marginal point).' Personal communication, 31 May 2002.

32 In Marrakech, Norway called for a more formal cooperation between the CBD and UNFCCC in order to avoid negative effects from forestry activities under the Kyoto Protocol ('Submission from Norway').

33 Another important biodiversity instrument is the EU Birds Directive (79/409/EEC).

34 Ministry of Environment, Budget 2002, http://odin.dep.no/bud2002/sb/md/kap09.htm.

[35] *Final Report on Financing Natura 2000*, Working Group on Article 8 of the Habitat Directive (joint working group of member state and European Commission experts) Brussels, November 2002.

[36] In Case C-103/00, Commission of the European Communities v Hellenic Republic. THE COURT hereby: 1. 'Declares that by failing to take, within the prescribed time-limit, the requisite measures to establish and implement an effective system of strict protection for the sea turtle Caretta caretta on Zakinthos so as to avoid any disturbance of the species during its breeding period and any activity which might bring about deterioration or destruction of its breeding sites, the Hellenic Republic has failed to fulfil its obligations under Article 12(1)(b) and (d) of Council Directive 92/ 43/EEC of 21 May 1992 on the conservation of natural habitats and of wild fauna and flora; 2. Orders the Hellenic Republic to pay the costs'.

[37] Personal communication, November 2001, Gunn Paulsen, DN.

[38] Personal communication, NN, Ministry of the Environment, 3 September 2001. This view was substantiated by personal communication, Ministry of Agriculture, 16 April 2002, in the sense that the Ministry of Agriculture prefers setting political goals for forest conservation rather than trying to determine scientific goals.

[39] The economic interests tied to fish farming are vast, and growing and strong criticism has been raised about the scientific background for the conservation plans (Kaarbø, 2001; Moy, 2001).

[40] The environmental departments at the county level repeatedly draw the shorter straw in confrontations with the Fisheries Directorate (Olsen, 2001).

[41] Compared to the government subsidies provided for building forest roads and accessing timber in remote and difficult areas, only one forth is granted to environmental measures in the forest sector (see Appendix 1).

[42] See also SABIMA, (2003).

[43] Note from from Nature and Youth (Natur & Ungdom) to the St.prop. 1, National Budget 2003 (Statsbudsjettet 2003). Also personal communication with central NGO leader, 24 May, 2002.

[44] See also Barth-Heyerdahl (2002) and Håpnes (2002).

[45] Samarbeidsrådet for biologisk mangfold; Cooperative Board for Biodiversity, including a large number of NGOs and scientific organisations.

[46] The Intergovernmental Panel on Forests (IPF) was established by the Commission on Sustainable Development and led to the UN Forum on Forests (UNFF).

[47] According to a source in the Ministry of Environment, the forest department of the Ministry of Agriculture is 'disrupting' the biodiversity issue – they represent a different culture and quite different priorities (personal communication, Ministry of the Environment, 3 September 2001).

[48] This section builds partly on an interview with Anne Marie Skjold, the Ministry of Foreign Affairs (20 January 2000) and partly on the author's own observations during the IFF-4 meeting (New York, January 2000). See also Rosendal (2001a).

[49] This conflict appeared early in the aftermath of the UNCED. The author was at that time part of an interministerial (Ministry of the Environment and Ministry of Agriculture) working group on the White Paper to Parliament on the UN Conference on Environment and Development in Rio de Janeiro. At these meetings, the forest department of the Ministry of Agriculture was present, and the group had great difficulties in reaching agreement on formulations pertaining to the forest sector.

50 The legal principle of 'freedom with responsibility' pertaining to the environmental obligations of forest owners has to some extent backfired, as it also implies no economic compensation. Personal communication, Arne Ivar Sletnes (Ministry of Agriculture) and Olav Bakken Jenssen (Ministry of the Environment), April, 2002.

51 Likewise, the emerging patent regimes, such as the EU Patent Directive and the TRIPS Agreement may possibly, but not necessarily, have the potential to work against the biodiversity objectives. The EU Patent Directive may be accepted with or without accompanying legislation to enhance benefit sharing with the countries providing genetic resources.

52 This sentiment was clearly expressed during the '*Consensus Conference on Management of Biological Diversity in Norway*', 21-22 May 1996, organised by the Research Council of Norway.

53 As pointed out in interview with Olav Bakken Jensen (Ministry of the Environment), and Arne Ivar Sletnes (Ministry of Agriculture), April 2002.

References

Aalde, O., Ekanger, I., Rosland, A., Grunne, T., Sollie, P.V., Torssen, H. and Lindestad, B.H. (1997), *Skog og klima. Skog og treprodukters potensiale for å motvirke klimaendringer.* Working group for climate and forestry measures, created in April 1996, report to the Norwegian Ministry of Agriculture, December, Oslo.

Barth-Heyerdahl, L. (2002), 'Skogvern til 3.5 mrd', *Natur & Miljø Bulletin*, no. 8, p. 5.

Bjørnes, T. and Lafferty, W. (2000), *Miljøvernlederstillinger og Lokal Agenda 21. Hva er status?* Report 1/2000, SUM, University of Oslo.

de Wit, H. and Kvindesland, S. (1999), *Carbon Stocks in Norwegian Forest Soils and Effects of Forest Management on Carbon Storage*, Rapport fra skogforskningen – Supplement 14, pp. 1-52, Norwegian contribution to the IPCC Special Report on Land Use, Land Use Change and Forestry.

Eikeland, P.O. (forthcoming) 'Environmental Innovation in the Norwegian and Swedish Electricity Supply Industry' forthcoming doctoral thesis.

Ernst & Young (2001), *Focus on Fundamentals. The Biotechnology Report. Ernst & Youngs 15th Annual Review*, Ernst & Young, October.

Fauchald, O.K. (2001), 'Patenter og allmenningens tragedie', *Lov og Rett, Norsk juridisk tidsskrift*, 40. årgang.

Fiske, P., Lund, R.A., Østborg, G.M. and Fløystad, L. (2001), *Escapees of Reared Salmon i Coastal and Marine Fisheries in the Period 1989-2000* (Rømt oppdrettslaks i sjø- og elvefisket i årene 1989-2000), NINA Oppdragsmelding 704, Norwegian Institute for Nature Research, Trondheim pp. 1-26.

Fleming, I.A., Hindar, K., Mjølnerød, I.B., Jonsson, B., Balstad, T. and Lamberg, A. (2000), 'Lifetime Success and Interactions of Farm Salmon Invading a Native Population', *Proc. R. Soc. Lond.*, vol. 267, pp. 1517-23.

Gleckman, H. (1995), 'Transnational Corporations' Strategic Responses to "Sustainable Development"', in H.O. Bergesen and G. Parmann (eds), *Green Globe Yearbook of International Co-operation on Environment and Development 1995*, The Fridtjof Nansen Institute and Oxford University Press, Oxford, pp. 93-106.

Gulbrandsen, L.H. (2001), *Konvensjonen om biologisk mangfold og norsk skogbruk. Strategisk, normativ eller ingen tilpasning*? FNI Report No. 17/2001, The Fridtjof Nansen Institute, Lysaker.

Gulbrandsen, L.H. (2003), 'The Evolving Forest Regime and Domestic Actors: Strategic or Normative Adaptation?', *Environmental Politics*, vol. 12, no. 2, pp. 95-114.

Hindar, K. (2001), *Interactions of Cultured and Wild Species*, Marine Aquaculture and the Environment, University of Massachusetts, Boston, MA, January 11-13.

Hindar, K, Ryman, N. and Utter, F. (1991), 'Genetic Effects of Cultured Fish on Natural Fish Populations', *Can. J. Fish. Aquat. Sci.*, vol. 48, pp. 945-57.

Hustad, H. (2002), *EU's habitatdirektiv og norsk naturforvaltning*, forundersøkelse (preliminary study) March, WWF-Norway, Oslo.

Håpnes, A. (2002), 'Forskerne er enige: Mer skogvern!', *Verdens natur*, no. 3, WWF-Norway, pp. 8-9.

Johnston, S. and Yamin F. (1997), 'Intellectual Property Rights and Access to Genetic Resources' in J. Mugabe, C.V. Barber, G. Henne, L. Glowka and A. La Viña (eds), *Access to Genetic Resources: Strategies for Sharing Benefits*, ACTS Press, Nairobi.

Kaarbø, A. (2001), 'Ny regjering må legge vekk omstridt plan om laksevern', *Aftenposten*, 14 October, p. 3.

Koester, V. (1997), 'The Biodiversity Convention Negotiation Process and Some Comments on the Outcome', *Environmental Policy and Law*, vol. 27, no. 3, pp. 175-92.

Kremen, C., Niles, J.O., Dalton, M.G., Daily, G.C., Ehrlich, P.R., Fay, J.P., Grewal, D. and. Guillery, R.P (2000), 'Economic Incentives for Rain Forest Conservation Across Scales', *Science*, vol. 288, no. 5472, pp. 1828-32.

Ministry of Agriculture (1989), *Versitile Forestry. The Relationship between Forestry, the Natural Environment and Outdoor Life* (Flersidig skogbruk. Skogbrukets forhold til naturmiljø og friluftsliv), Report No. NOU 1989:10, Ministry of Agriculture, Oslo.

Ministry of Agriculture (1999), *Verdiskapning og miljømuligheter i skogsektoren (Skogmeldingen)*, Report No. 17 to the Storting (1998-99), Ministry of Agriculture, Oslo.

Ministry of the Environment (1981), *Protection of Norwegian Nature* (Vern av norsk natur), Report No. 68 to the Storting (1980-81), Ministry of the Environment, Oslo.

Ministry of the Environment (1989), *Environment and Development. Programme for Norway's Follow-up of the Report of the World Commission on Environment and Development*, (Miljø og Utvikling. Norges oppfølging av Verdenskommisjonens rapport), Report No. 46 to the Storting (1988-89), Ministry of the Environment, Oslo.

Ministry of the Environment (1991), *Report on Large Predators* (Rovviltmeldingen), Report No. 27 to the Storting(1991-1992), Ministry of the Environment, Oslo.

Ministry of the Environment (1996), *Environmental Protection Policies for Sustainable Development* (Miljøvernpolitikk for en bærekraftig utvikling), Report No. 58 to the Storting (1996-97), Ministry of the Environment, Oslo.

Ministry of the Environment (1999), *Til laks åt alle kan ingen gjera? Om årsaker til nedgangen i de norske villaksbestandene og forslag til strategier og tiltak for å bedre situasjonen*, Report No. NOU 1999:9, Study commissioned by royal declaration on 18 July 1997 and delivered to the Ministry of the Environment on 12 March (Utredning fra et utvalg oppnevnt ved kongelig resolusjon av 18. juli 1997. Avgitt til Miljøverndepartementet 12. mars), Ministry of the Environment, Oslo.

Ministry of the Environment (2001), *Biological Diversity. Sector Responsibility and Coordination* (Biologisk mangfold. Sektoransvar og samordning), Report No. 42 to the Storting (2000-2001), Ministry of the Environment, Oslo.

Ministry of the Environment (2002), *Government Environmental Protection Policies and the Environmental Status of the Nation* (Regjeringens miljøvernpolitikk og rikets miljøtilstand), Report No. 25 to the Storting (2002-2003), Ministry of the Environment, Oslo.

Ministry of Foreign Affairs (1988), *Norwegian Contributions to International Activities for Sustainable Development* (the 'Blue Book'), Ministry of Foreign Affairs, Oslo.

Ministry of Foreign Affairs (1995), *Evaluation Report. Integration of Environmental Concerns into Norwegian Bilateral Development Assistance: Policies and Performance,* an evaluation by the Fridtjof Nansen Institute and ECON Centre for Economic Analysis, Report No. 5:1995, Ministry of Foreign Affairs, Oslo.

Moy, R. (2001), 'Fiskeoppdrettere kritisk til lakseforskning: Krever full offentlig granskning', *Aftenposten*, 16 October, p. 36.

NINA and Skogforsk (2002), *Evaluation of Forest Protection in Norway* (Evaluering av skogvernet i Norge), Norwegian Institute for Nature Research, Trondheim, and Norwegian Forest Research Institute (Skogforsk), Ås.

OECD (1993), *OECD Environmental Performance Reviews. Norway*, OECD, Paris.

OECD (2001a), *Environmental Performance Reviews. Achievements in OECD Countries*, OECD, Paris.

OECD (2001b), *OECD Environmental Performance Reviews. Norway*, OECD, Paris.

Olsen, K. (2001), 'Lite miljøhensyn i oppdrettssaker', *Aftenposten*, 20 August, p. 28.

Porter, G. (1993), 'The United States and the Biodiversity Convention', *International Conference on the Convention on Biological Diversity*, African Centre for Technology Studies, Nairobi, Kenya, January 26-29.

Quammen, D. (1996), *The Song of the Dodo. Island Biogeography in an Age of Extinctions*, Random House, London.

Raustiala, K. (1997), 'Global Biodiversity Protection in the United Kingdom and the United States', in M.A. Schreurs and E.C. Economy (eds), *The Internationalization of Environmental Protection*, Cambridge University Press, Cambridge, pp. 42-73.

Reinvang, R. (2003), *Norway, EU, and Environmental Politics*, Report 1/03, Framtiden i Våre Hender, Oslo.

Rosendal, G.K. (1994), 'Implications of the US "No" in Rio', in V. Sánchez and C. Juma (eds), *Biodiplomacy: Genetic Resources and International Relations*, African Centre for Technology Studies, ACTS Press, Nairobi, pp. 87-103.

Rosendal, G.K. (2000), *The Convention on Biological Diversity and Developing Countries*, Kluwer Academic Publishers, Dordrecht.

Rosendal, G.K. (2001a), 'Overlapping International Regimes: The Forum on Forests (IFF) between Climate Change and Biodiversity', *International Environmental Agreements: Politics, Law and Economics*, vol. 1, no. 4, pp. 447-68.

Rosendal, G.K. (2001b), 'Impacts of Overlapping International Regimes: The Case of Biodiversity', *Global Governance*, vol. 7, no. 1, pp. 95-117.

Rosendal, G.K. (2001c), 'Institutional Interaction in Conservation and Management of Biodiversity: An Inventory', retrieved from www.ecologic.de/english/interaction/.

SABIMA (Samarbeidsrådet for biologisk mangfold) (2003), *Forskningsetiske aspekter vedrørende prosjektet 'Miljøregistreringer i skog'*. (Aspects of research ethics with respect to the project 'Environmental Registrations in Forestry – Biodiversity'), Note, 11 June, retrieved 5 February 2004 from www.sabima.no/index.htm.

Schei, P.J. (1997), 'Konvensjonen om biologisk mangfold' in W.M. Lafferty, O.S. Langhelle, P. Mugaas and M. Holmboe Ruge (eds), *Rio + 5: Norges oppfølging av FN-konferansen om miljø og utvikling*, Tano Aschehoug, Oslo, pp. 244-58.

Solberg, B. (1997), 'Forest Biomass as Carbon Sink – Economic Value and Management/Policy Implications', *Meddelelser fra Skogforsk*, no. 48, pp. 373-85.

Solås, Asbjørn (1998), 'Feil om CO_2 tiltak i skog?', *Biolog*, vol. 16, no. 3-4, pp. 53-55.

Solås, A. and Aanderaa, R. (1998), 'Kombiner CO_2-lagring og biologisk mangfold', *Norsk Skogbruk*, no. 2, pp. 24-5.

Stokke, O.S. and Thommessen, Ø.B. (2002), *Yearbook of International Co-operation on Environment and Development 2002/2003*, Earthscan, London and Sterling, VA.

Svensson, U. (1993), 'The Convention on Biodiversity – A New Approach', in S. Sjøstedt and H. Aniansson (eds), *International Environmental Negotiations*; *Process, Issues and Context*, FRN Utrikespolitiska Institutet, Stockholm, pp. 164-91.

Ulfstein, G. (2001), 'Bernkonvensjonen som rettslig ramme for norsk rovdyrpolitikk', unpublished manuscript, Oslo University, 18 June.

UNDP (1999 and 2000), *Human Development Report*, United Nations Development Programme, New York, Chapter 2.

UNEP (1995), *Global Biodiversity Assessment*, Cambridge University Press, Cambridge.

Wilson, E.O. (ed) (1988), *Biodiversity*, National Academy Press, Washington D.C.

World Commission on Environment and Development (1987), *Our Common Future*, Oxford University Press, Oxford.

WRI (1991), *Minding the Carbon Store: Weighing US Forestry Strategies to Slow Global Warming*, Mark C.Trexler, World Resources Institute,Washington, DC.

Appendix 1 National budgets, 2002-1990, St prp No. 1. Selected posts on biodiversity conservation and forest management in the Ministry of Environment and the Ministry of Agriculture (all in million NOK)

	2002	2001	2000	1999	1998	1997	1996	1995	1994	1993	1992	1991	1990
1. Protected areas (county level)	17.7 (39.6)[a]	18.5 (78)	19.5 (37)	13.5	10	24 (22)	24	-	-	-	-	-	-
2. Protected areas (state level)	30.9	22.7	12.7	14.7	7.9	(6)	14.7	111.3	69	57	48	46.9	49.9
3. Coniferous Forests	70.9[b] (230)	30 (174)	26 (65)	26	35	45	50	(10)	(22)	(180)	(35)	(28)	-
4. Compensation – predators	83	80	80	96	70	43	36	25.6	24	16	18	16.4	11.4
5. Preventive measures – predators	34.5	40	28	-	-	-	-	-	-	-	-	-	-
6. Forest roads etc. – forest sector	100.4	107.9	128	123.9	107.7	90	89.3	126.7	149.6	148	192	120 + 67	108 + 93
7. Environmental measures – forest sector	23.4	23	23	22.6	26.3	24.6	14.5	14	14	-	-	-	-

[a] All figures in parenthesis indicate the total amount endorsed by the Parliament (spesial- eller tilleggsfullmakt).

[b] In the budget for 2003 this is increased to 110 million, which will be used to approach the goal of one per cent conservation in the Protection Plan for Coniferous Forests.

Sources:

1. Ministry of Environment, Chapter 1427, Post 32: Protected areas, county level (compensation and restraints). Since 1996.
2. Ministry of Environment, Chapter 1427, Post 30: Protected areas, state level (compensation and restraints).
3. Ministry of Environment, Chapter 1427, Post 33: Protection of Coniferous Forests. Since 1996.
4. Ministry of Environment, Chapter 1427, Post 72: Compensation relating to damage caused by wild predators.
5. Ministry of Environment, Chapter 1427, Post 73: Preventive measures and readjustments relating to damage caused by wild predators. Subsumed under post 72 until 1999.
6. Ministry of Agriculture, Chapter 1142, Post 71: Forest sector developments and road building.
7. Ministry of Agriculture, Chapter 1142, Post 76: Resource and environmental instruments for the forest sector.

Chapter 9

Comparative Analysis and Conclusions

Jon Birger Skjærseth, Steinar Andresen, Hans-Einar Lundli, Tom Næss, Marit Reitan, G. Kristin Rosendal and Jørgen Wettestad

We now shift focus from the study of single cases to comparative patterns. In Chapters 3-8, we have analysed the extent to which and how Norway has achieved its goals in various issue areas of international environmental cooperation. Since target groups are located both inside and outside Norway's borders, two dimensions of goal attainment have been evaluated: international and domestic. The aim of this chapter is to take a step back and look for patterns in Norwegian goal attainment at both international and domestic levels. We focus particularly on how institutional factors have enabled or constrained goal attainment within various issue areas at both the international and domestic levels. More specifically, we considered the particular roles of the characteristics of the core regime, linked regimes and institutionalization of foreign and domestic environmental policy (see Chapter 1 for a more detailed description of the research questions). The institutional approach is supplemented with an alternative explanatory approach that looks at problem type, as defined by cost-benefit distribution and configuration of actor interests.

The conventional approach to studying regime effectiveness is to look at how international regimes influence and engage member states in the objectives of the regime. This book has taken the opposite point of departure by looking at how one state engages a web of international regimes to pursue national goals at both international and national levels. The literature on regime linkages has tended to emphasize problems caused by regime congestion and density. The analysis provided below shows, however, that one state with a coherent environmental policy will mainly benefit environmentally from the crossfire of a high and growing number of international environmental regimes.

Norwegian Goal Attainment

To what extent has Norway been able to attain its environmental goals internationally and domestically? In Chapter 2, we distinguished between what *can* be achieved and what *has* been achieved. What *has* been achieved has first been analysed in terms of international goal attainment as measured by the match between Norwegian positions and regime objectives, and wherever feasible, change in the behaviour of foreign target groups. Second, we have measured domestic goal attainment in terms of change in the behaviour of domestic target groups, such as industry, agriculture and municipalities.

Our assessment of what *can* be achieved is based on the criterion of explicit national environmental goals. Before we present the results, however, some notes of caution are in order. The first challenge in the comparative analysis is the problem of vagueness. In some cases, goals have been framed in a precise way that leaves little room for interpretations. In other cases, goals have been formulated more ambiguously. With regard to domestic goal attainment, ozone depletion as well as air- and marine pollution are characterised by clear percentage goals linked to deadlines and baselines. In the case of biodiversity, national goals cover a broad field of activities, and the goals vary significantly in specificity. By implication, the chapter on biodiversity has not aimed at evaluating all aspects pertaining to biodiversity policies in Norway. Instead, it focuses mainly on conservation and management of forests as well as habitat conservation and management of wild salmon, for both substantial and methodological reasons (see Chapter 8). A different problem has surfaced in the case of climate change (see Chapter 7). Here it has been difficult to distinguish between international and domestic goals, since the two are closely intertwined: domestic targets have partly been conditioned upon the adoption of a binding climate treaty based on international flexibility mechanisms. Moreover, the unilateral stabilisation goal was replaced by the national commitments agreed upon in Kyoto.[1] We have chosen to base our evaluation of the climate case on a generous interpretation of Norway's stabilisation target. A high degree of domestic goal attainment equals a significant reduction in *the expected increase* in carbon dioxide (CO_2) emissions from 1990 to 2000, compared to the 1990 baseline.

The second challenge relates to changes in goals over time. With the exception of the whaling case, the goals selected in this study have been extracted from the Norwegian 1988-89 White Paper on sustainable development. The whaling issue was not included in the White Paper because it was regarded as a question of managing marine living resources rather than an environmental issue, so our starting point has been Norway's 1986 decision to halt commercial whaling. Thus because the goals for all of the issue areas covered in this study were set around the same time, i.e., in the late 1980s, general fluctuations in societal demands for environmental improvement and change in the strength of the green movement are unlikely to account for *differences* in goal attainment between various issue areas.

A third challenge is related to aggregation of goal attainment when there are several different regime components and time phases. In some cases, such as air pollution, Norway's achievements vary significantly between different substances covered by separate international protocols. For practical reasons, we simply used a score that reflects the average for all the regime components. However, we have disaggregated the separate substances in the explanatory section to make the conclusions more robust. With regard to time phases, we compare domestic goal attainment according to deadlines included in national goals or international obligations to which Norway has voluntarily agreed. With the above caveats, we present the comparative results in Table 9.1.

Table 9.1 Norwegian goal attainment internationally and domestically

	Whaling	Ozone	Bio-diversity	Air pollution	Cli-mate	Marine Pollution
Goal attainment internationally	M[c]	H	H	H[d]	H	H[c]
Goal attainment domestically	–[a]	H	M[d]	L/M[d]	L[b]	M[d]

H: High; M: Medium; L: Low. Measured in relative and absolute terms.

[a] Goal attainment in whaling is almost exclusively a question of influencing foreign target groups.

[b] Goal attainment in climate is based on the interpretation above.

[c] Significant changes in goal attainment have occurred over time.

[d] Significant differences in goal attainment between various components or substances.

We can immediately see that Norway has been generally more successful 'abroad' than at 'home'. Norway has been in the forefront, pushing for stringent international commitments, or acting as a mediator in cases where developing countries have played an important role. This behaviour, clearly in the spirit of the Brundtland Commission, is highly consistent with the main lines and principles underlying Norway's environmental policy (see Chapter 1).

On the other hand, Norway has a relatively poor record (compared to goal attainment internationally) when it comes to domestic goal attainment, with the exception of ozone depletion. In most cases, Norway has faced difficulties at home: municipalities, agriculture, forestry, transport, energy and to some extent industry have not delivered results in accordance with stated objectives. These problems have also led Norway to breach international agreements – as with emissions of volatile organic compounds (VOCs), where actual achievements lag far behind international obligations (see Chapter 5). Consequently, Norway is at odds with its principles of putting its own house in order and complying with international commitments. The upshot of these patterns is simply that Norway has performed better with regard to foreign than domestic target groups.

The following discussion recapitulates and evaluates goal attainment for each case.

Goal Attainment Internationally

The cases are evaluated in terms of the congruence between Norwegian positions and regime objectives. Wherever feasible, we take one step further to assess change in the behaviour of target groups that are located outside Norway.

The whaling case displays significant change in goal-achievement over time (see Chapter 3). This case deviates from the others since target groups have primarily been located outside Norway. In order to hunt whales, which has been and still is the goal, Norway has directed its whaling policy towards other states and non-state actors that oppose whaling. In the mid 1980s, Norway was forced by the parties to the International Whaling Commission (IWC) and the US to stop commercial whaling. In 1993, however, commercial whaling was resumed, and export of whale products was allowed from 2001. While the resumed harvest of minke whales indicates an increase in goal attainment, the moderate Norwegian catch appears to be a concession to the anti-whaling forces. This points to a 'medium' score on goal attainment (see Table 9.1).

The ozone case is comparable with whaling in the sense that the major target groups – the chemical companies that produce ozone-depleting substances (ODS) – are located outside Norway (see Chapter 4). Unlike the whaling case, however, Norway has not fought against a strong ozone regime clashing with Norwegian interests, but rather has supported stringent regime objectives in order to achieve its own ozone policy goals. Together with the Toronto group of states and later the EU, Norway has pushed consistently and successfully for global reductions of ODS. This effort has proved successful in the sense that the Montreal Protocol has contributed to nearly a total phase-out of ODS in OECD countries.

Norway's air pollution policy has aimed at reducing transboundary emissions that contribute to acidification and related environmental damages in Norway (see Chapter 5). The principal instrument has been the 1979 Convention on Long-Range Transboundary Air Pollution (CLRTAP). As a net importer of air pollution, Norway (together with Sweden) was one of the main architects behind the 1979 agreement, which has developed into various international protocols on specific substances. Seen from a Norwegian perspective, goal attainment internationally has been high. From 1980 to 1997, for example, European emissions of sulphur dioxide (SO_2) were reduced by 60 per cent, and sulphur deposition over Norway has been halved. Note that Norway's position in the international negotiations has varied according to the substances in question: Norway pushed for stringent commitments in the case of SO_2, but was more reluctant with regard to other substances such as VOCs and nitrous oxide (NO_x). Still, there is generally a high overall match between Norway's positions and CLRTAP objectives.

The case of marine pollution is quite similar to that of air pollution (see Chapter 6). As a net importer of marine pollution resulting from the counter-clockwise direction of the North Sea currents, Norway has supported a strong regime on marine pollution. Norway took the initiative to establish the first international convention on the protection of the North-East Atlantic, the 1972 Oslo Convention on dumping and incineration at sea. Norwegian goal attainment internationally has been high: dumping and incineration have been phased out, and far-reaching joint commitments on hazardous substances and nutrients have been followed by significant reductions of controlled substances in most North Sea states.

Ever since the late 1980s Norway has been working for an international climate regime with binding emission commitments (see Chapter 7). This ambition is expected to materialize if Russia ratifies the Kyoto Protocol. Several specific climate-regime arrangements and principles have been important to Norway because of its unusual role of being an oil-producing nation and yet already depending mostly on clean, renewable energy through hydroelectric power (which increases Norway's marginal abatement costs significantly). There is a considerable match between Norway's positions and the actual outcomes of climate negotiations. First, Norway worked hard to include differentiated commitments for industrialized countries, which has become an integral part of the Kyoto Protocol. While the Kyoto Protocol commits the industrial countries to reducing their greenhouse gases by 5.2 per cent collectively, Norway is allowed to increase its emissions by one per cent from 1990 to 2008-12. Second, Norway argued for a comprehensive approach that would allow parties to reduce greenhouse gases other than CO_2, and the Kyoto Protocol currently allows parties to reduce any of six greenhouse gases measured in CO_2 equivalents. Third, Norway supported the idea of joint implementation and emission trading between countries, in line with the principle of cost effectiveness. Along with the Clean Development Mechanism (CDM) these flexibility mechanisms represent the core of the Kyoto Protocol. Finally, Norway has been arguing for an international climate regime based on binding emission commitments in line with the Kyoto Protocol. Taken together, these four points show a high overall match between Norway's positions and the various elements of the climate regime.

In the case of biodiversity policy (see Chapter 8), Norway displayed a high profile in the international negotiations on the 1992 Convention on Biological Diversity (CBD). The three-fold package of objectives included in the CBD fits well with the position taken by Norway: to ensure both conservation and sustainable use of biological diversity, as well as equitable sharing of benefits rising from the use of genetic resources. More specifically, Norway was pleased with the principle that all countries have a common responsibility to share the costs of conservation. Other central Norwegian goals were the principles that the CBD should include all biological diversity, wild and domesticated, and that it should pertain to sustainable use along with conservation.

We may conclude that there has been a high overall match between Norwegian positions in relation to the core regimes and regime objectives. In most cases, Norway has pushed for strong regimes that have increased the prospects for influencing foreign target groups – as shown clearly in the ozone regime, which has contributed to a significant reduction in ODS. Whaling, however, constitutes a deviant case since the Norwegian position has been at odds with the main objective of the IWC. Nevertheless, Norway has succeeded in increasing the level of its goal attainment from the time it was forced to halt whaling.

Goal Attainment Domestically

Domestic goal attainment has been measured in terms of the congruence between environmental policy goals (often derived from international obligations) and the behavioural change of relevant target groups. As noted, Norway's whaling policy is primarily about foreign policy-making rather than domestic implementation. However, the decision made by the government to resume export of whale products has made the question of compliance more relevant. Previous suspicions of illegal trade with minke whale products has necessitated strict control requirements based on DNA testing of every whale taken by Norwegian whalers. The system also requires that the importers have a similar DNA system in place. This control system appears promising, but it is too early to judge how it will work in practice.

Domestic implementation with regard to ozone depletion has been an outright success. The ambitious domestic goals on ODS consumption formulated in the late 1980s have been achieved. Indeed, actual achievements have exceeded original goals, since domestic consumption of ODS has been reduced by 98 per cent compared with the 1986 level.

Norwegian air pollution policy has been more a failure than a success compared to stated Norwegian goals. Surely, transboundary air pollution has been significantly reduced, but Norway has experienced a number of problems regarding domestic goal attainment. Norway's performance on SO_2 has clearly been a success (75 per cent reduction between 1980 and 1997), but Norway has experienced significant problems regarding emissions of NO_x and VOCs. The results on VOCs have been described as an outright failure: Norway aimed to stabilise overall emissions by 1999 at 1989 levels, and to reduce emissions by 30 per cent within country-specific geographic areas. Instead, emissions have actually increased by around 20 per cent from the 1989 level. With regard to air-quality goals established in 1997 and 1998, goal attainment is somewhat mixed, but generally not satisfactory.

In the case of marine pollution, Norway's goal attainment is mixed. On one hand, many regulated hazardous substances have been reduced significantly and emissions of phosphorous substances have been cut by 50 per cent in sensitive areas in line with stated goals. On the other hand, Norway has not even come close to reaching its goal on nitrogenous substances, and there are indications that the reductions in hazardous substances have stagnated or have been going in the wrong direction for some substances.

With climate policy, Norway initially aimed at stabilizing emissions of CO_2 between 1989 and 2000. Later, it adopted the relatively less stringent Kyoto target. However, even with a less ambitious goal it has had little success in reducing the expected increase in domestic CO_2 emissions. The White Paper following up the Brundtland Report estimated that emissions would increase by 28 per cent between 1989 and 2000 if no measures were adopted. Despite regulations, including carbon taxes, actual emissions were 20 per cent higher in 2000 than in 1989 – only a modest eight per cent better than business as usual.

Norway kept a high profile in the international negotiations on biodiversity, but goal attainment at home has not been impressive. First, the magnitude of forest management and protection deviates significantly from stated goals. Norway has protected only one per cent of the productive forest, as compared to a scientific recommendation of five per cent. In addition, Norway spends only a fraction of that spent by e.g. Sweden and Finland on forest conservation. Second, species management has fallen short of ambitions, particularly related to the management of wild salmon. In the late 1980s, Norway aimed to establish 'fish-farm-free zones' along the coast in order to protect the wild populations of salmon. It took until 2003 to decide on the establishment of these zones, and by then the protection regime associated with them had been both diluted and diminished. However, the situation is somewhat better with regard to habitat conservation. Protected areas currently add up to about ten per cent, which compares favourably with the ten per cent recommended by the OECD.

We may conclude that there has been a relatively low overall match between Norway's goals and achievements on the ground, i.e. among the domestic target groups causing the problems in the first place. In most cases, Norway has faced significant problems in implementing measures to achieve stated goals. The clearest exception is ozone depletion, where the country's goal attainment in terms of ODS consumption has exceeded initial goals.

Explanatory Factors: International and Domestic Institutions

In this section, we have narrowed our focus to institutional 'attributes' at domestic and international levels by combining the study of international and domestic institutions. The challenges faced by Norway in reaching its goals correspond with the location of target groups. In most cases, Norwegian goal attainment has depended on the government's ability to influence national *and* foreign target groups. If target groups are mainly located abroad, as in the whaling case, goal attainment depends on successful foreign policy 'upstream' directed at relevant regimes. The question then becomes, To what extent, and how, have institutional factors affected Norway's high level of international goal attainment? If target groups are mainly located at home, goal attainment depends on successful implementation of national policy 'downstream' directed towards domestic target groups, so the main question becomes: To what extent and how have institutional factors determined Norway's mixed level of domestic goal attainment?

The Upstream Process: Influencing Outsiders

Goal attainment internationally is institutionally considered to be a function of three main determinants. First, successful goal attainment depends on the extent to which there is inter-ministerial coordination of positions to frame a *coherent* foreign environmental policy. Case-study authors have explored two distinct qualities of domestic institutions likely to affect the level of coherency: distribution of competence between ministries, and the number and types of ministries and agencies involved in shaping foreign environmental policy. Second, Norwegian goal attainment depends upon the *means* by which Norway has exercised its influence, as well as the *receptivity* and *strength* of the core regime. Means of influence points to the level of activity in the international negotiations and different types of leadership applied to affect core regimes. Degree of receptivity determines the scope for influencing the regime in question, whereas the strength of the regime determines the impact of the regime on target groups outside Norway. The authors of the case studies presented here have explored various regime qualities and sources of influence that are likely to affect goal attainment internationally. Third, we have explored the extent to which and how other functionally *linked* regimes have conformed with the objectives of the core regimes and Norwegian positions, on the assumption that a high match in objectives or good opportunities for 'venue shopping' will facilitate Norwegian goal attainment internationally. The importance of other states for Norway's goal attainment internationally is dealt with towards the end of this chapter.

Domestic institutions: *coherent foreign policy.* Three of the cases display a high degree of coherency between the ministries and agencies involved, whereas some disagreement has been reported in two cases: biodiversity and climate change.[2] The whaling case deviates from this pattern in the sense that significant disagreement was been observed initially, although it changed towards consensus over time. The level of coherence witnessed was assumed to be closely related to the number and type of ministries and agencies involved in shaping Norwegian positions. In addition, we assumed that coherence would be strengthened in cases where one ministry was in charge of formulating national positions in international negotiations.

A high level of coherence goes hand in hand with a relatively low and stable number of participating ministries across the regimes in which Norway participates. In most cases, between two and four ministries have been directly involved in shaping the positions taken by Norway. The typical constellation of governmental actors involved has been a triad: the Ministry of the Environment, the Ministry of Finance, and the Ministry of Foreign Affairs. One ministry has been in charge in five out of the six cases. In four cases, the Ministry of the Environment has had responsibility for the international negotiations. More important is perhaps the observation that not only does the same ministry 'represent' environmental concerns, but a small group of key persons in the Ministry of the Environment has been in charge of different negotiations.[3] This is likely to improve coordination of positions across various regimes. In the case of climate change, competence has been split between the Ministry of Foreign Affairs and the Ministry of the Environment.

Let us now take a closer look at those cases where disagreement on Norwegian positions has been reported. The whaling case is particularly interesting here since goal attainment is essentially a matter of influencing foreign target groups (see Chapter 3). In the 1980s, the ministries of foreign affairs, trade, and the environment became quite receptive to the external anti-whaling forces, which threatened Norway with trade sanctions and consumer boycotts. These ministries were in turn increasingly supported by the green movement, the media and Norwegian export industries. The Ministry of Foreign Affairs, which formally appoints Norway's IWC Commissioner, leaned toward the position that Norway should not resume commercial whaling – a position based on the perception that a pro-whaling position could threaten Norway's reputation as a true 'internationalist'. On the other hand, the Ministry of Fisheries, supported by fishers and whalers, was strongly in favour of resumption.

The whaling dispute became elevated to the top political level, and the Prime Minister, Gro Harlem Brundtland, and the Prime Minister's Office intervened by launching a two-step strategy. First, a scientific strategy was initiated based on independent research (i.e. independent of Norwegian interests) in order to increase legitimacy. The scientific strategy proved successful when the IWC Scientific Committee endorsed the Norwegian estimate of minke whale stocks. Second, an offensive political strategy was launched to persuade opponents at home and abroad that whaling could, and should as a matter of principle, be resumed irrespective of its marginal economic role. Even the green movement in Norway was persuaded to back the official position. In the end, all major Norwegian ministries and agencies and related interest groups rallied behind the official position, thus collectively framing a coherent foreign whaling policy. The whaling case shows how internal fragmentation in foreign policy can be overcome by means of governmental leadership.[4]

There have not been any serious coordination problems in the case of climate change despite the high number of sectors affected and the shared competence between the Ministry of the Environment and the Ministry of Foreign Affairs in shaping Norway's climate position (see Chapter 7). All ministries and subordinate agencies embraced the principle of cost-effectiveness in international agreements, and most of the specific Norwegian positions followed from this principle. Beneath the surface, however, the various ministries advocated different positions. The Ministry of the Environment placed relatively more emphasis on domestic cutbacks, while the Ministry of Finance strongly supported emission reduction abroad, fearing that domestic measures beyond the CO_2 tax would imply high costs. This difference in attitude towards the flexibility mechanisms can partly be traced back to different professions in the two ministries.

As noted, the Ministry of Fisheries played an important role in shaping the Norwegian position in the IWC. A similar development may be in progress with regard to biodiversity (see Chapter 8). The ministries of foreign affairs, the environment, and agriculture all participated in the Norwegian delegation during the negotiations on the 1992 CBD. At that time, the Ministry of Agriculture was not very active, and the ministries of foreign affairs and the environment were in complete agreement on the various Norwegian positions. Other ministries adopted a wait-and-see stance since they perceived that significant interests were not at stake. Negotiations on a forest convention that may weaken the CBD have threatened this harmony. In the aftermath of the 1992 UNCED, the Ministry of Agriculture succeeded in procuring a mandate allowing them to support a forest convention against the positions of the ministries of foreign affairs and the environment. The Ministry of Agriculture position was at odds with Norway's traditional role as a bridge-builder between developed and developing countries in the CBD. Accordingly, this case adds to the observation in the whaling case that sector ministries – fisheries and agriculture – can have significant influence on Norway's environmental foreign policy even if other ministries are in charge of the negotiations.

We may conclude that our main assumption has so far been supported, in the sense that a high level of coherence is accompanied by a high level of goal attainment internationally. In those cases where disagreement has been reported, different mechanisms have been at work in different cases. First, the *types* of ministries involved in developing Norwegian positions appear more important than the total number. Of particular interest is the observation that sector ministries representing fisheries and agriculture can play an important role in shaping environmental foreign policy even if other ministries are in charge. This probably reflects the traditionally strong position of these interests in Norwegian economy and politics, where rural interests have been important in a country that is sparsely populated. Second, we have found no indications that shared responsibility has created any problems. The main pattern shows that a high level of coherence has gone hand in hand with one specific ministry in charge: the Ministry of the Environment. Third, internal disagreement between ministries can be overcome through active governmental leadership. Leadership exercised at top political level reduced the level of internal disagreement among ministries in the whaling case, and contributed to an increase in Norwegian goal attainment.

Core regimes: means, receptivity and strength. Norway can influence the strength of international regime commitments that in turn can affect the behaviour of target groups abroad. Norway's goal attainment is thus conditional upon its own efforts (the efforts of others) and the receptivity and the strength of the core regime.

How has Norway exercised its influence in the core regimes? In Chapter 2, we suggested that negotiation skill and intellectual leadership based on scientific knowledge in specific issue areas were the most likely tools for small states in international environmental cooperation. Norway has played a very active role that comes close to leadership in the IWC and CLRTAP. In the IWC, Norway's influence was enhanced by increasing the legitimacy of the pro-whaling view by means of knowledge, persuasion and fair-play rather than tactics and threats. In the case of CLRTAP, the weight given to scientific knowledge facilitated Norwegian and Nordic influence due to substantial scientific/technical interests and competence in this area. In both cases, the scientific strategy exemplifies *intellectual leadership*.

In the other core regimes, Norway has also played an active role to further its goals mostly in close cooperation with the other Nordic states. Norway and the Nordic countries acted as mediators in the CBD negotiations – a role based more on ideal interests related to development aid principles than on economic interests or scientific knowledge. The developed countries possess the economic capacity and technology to benefit from the biological diversity of tropical areas in agribusiness and pharmaceutics, both of which are fields where Norway has had relatively limited industrial interests. Norway and the Nordic states participated actively in the negotiations leading to the ozone regime as part of a small group of 'pusher' states that had skilled negotiators that were able to persuade the EU, Japan and the former Soviet Union to take on emission reductions. The UNEP Secretariat also played a significant role in the CBD and ozone negotiations by supporting the position of developing countries and the Nordic position. In the case of climate change, Norway has participated actively all the way, and played a role as a mediator between the developed and developing countries. In the North Sea regime, Norway and the Nordic countries played a significant role based on marine scientific knowledge and political skill – the latter particularly with regard to dumping and incineration at sea. Norway took, as noted, the initiative to establish the first legal convention on dumping in the North-East Atlantic: the Oslo Convention.

The main conclusion is that Norway has played an active role to further its goals within all the core regimes selected in this study. This observation is in line with the OECD performance reviews. The first review noted that Norway had played a formidable role in international environmental politics (OECD, 1994, p.143), and, according to the most recent OECD review:

> ... Norway has played a *leading role* in the development of numerous regional environmental agreements. As one of the original proponents of sustainable development, it has also been *very active* in developing agreements on global environmental issues. (OECD, 2001, p.154, emphasis added)

The sources of Norway's activism lie in a combination of material and ideal interests. In the regional regimes on marine and air pollution, Norway is a net importer of pollution; accordingly, it has a clear interest in promoting strong international commitments. In the climate regime, Norway's energy-economic circumstances give it considerable economic interests, which point to differentiated emission targets and flexibility mechanisms. Norway has also supported the establishment of financial mechanisms that benefit developing countries. In the ozone and biodiversity regimes, limited economic interests were at stake initially, and Norway could serve as a mediator between developed and developing countries. In the IWC, Norway has mainly followed ideal interests linked to norms focusing on the 'right thing to do' for a small state in international politics.

A high level of international activity is in most cases necessary, albeit not sufficient, to attain international goals. How *receptive* the regime in question is will determine the extent to which efforts to influence regime objectives can translate into actual influence for 'pusher' states. Receptivity is here seen as a function of decision rules. Decision rules represent an ambiguous variable, as the extent to which one country would benefit from unanimity or majority decisions will depend upon that country's position in relation to the other parties in the regime. With this caveat in mind, we assumed that the influence of Norway would be *generally* strengthened when decision rules were based on consensus. This assumption was based on the premise that Norway's interests and positions would vary between the majority and the minority in different core regimes. Consensus allows countries to 'footnote' objections and to proceed by several 'fast-track' options. 'Footnoting' is important whenever a specific country wishes to reserve its position with regard to a (often more ambitious) majority, while 'fast-track options' allow a country to progress beyond the position of the minority.

Four of the core regimes looked at in this study are based on consensus, none on unanimity, and two (the ozone and whaling regimes) have provisions for majority decision rules. As expected, we found that consensus goes hand in hand with high receptivity and goal attainment internationally. In the North Sea regime, Norway has clearly benefited from various 'fast track' options. In this case, a change in decision rules from unanimity in practice to consensus and 'fast track' options significantly increased the influence of the 'pusher' states. In the case of air pollution, 'flexible consensus' facilitated Norwegian influence. The CLRTAP is a framework convention that has been developed and implemented by the adoption of specific protocols. Countries unwilling to accept a specific protocol simply did not sign. This mechanism ensured that reluctant parties could not block the progress of others.

The climate regime is particularly interesting with regard to decision rules. As noted in Chapter 2, Norway initially argued for majority decisions in order to strengthen the international authority on climate change. Paradoxically, if the most ambitious states (led by the EU) had got their will, the climate regime would have departed significantly from Norway's positions (see Chapter 7). In this case, the EU initially opposed the use of flexibility mechanisms and favoured equal emission targets based on three greenhouse gases only. The climate regime has, however, been based on consensus, with the less (compared to the EU) enthusiastic states (the US, Australia and Russia) determining the design of a climate regime largely in line with Norwegian positions. In negotiations on the Kyoto Protocol, 'fast-track' options like flexibility mechanisms and differential obligations for different states according to specific national energy-economic circumstances were crucial for Norway and for the agreement that was reached.

The whaling and ozone regimes further illustrate that majority decision rules can represent a double-edged sword seen from the perspective of a specific country. In 1989, the parties to the Montreal Protocol decided that decisions on all matters of substance should be taken by two-thirds majority vote, whereas decisions on matters of procedure should be taken by simple majority vote. Since Norway's position was in line with that of the majority, the majority decision procedure facilitated high goal attainment. In sharp contrast, the IWC regime has promoted significantly different objectives from those of Norway. Decision-making in the IWC is based on majority voting, so the anti-whaling coalition has been able simply to outvote Norway and the rest of the minority. The open access procedures led the anti-whaling coalition to recruit new states that strengthened the majority (see Chapter 3). However, anti-whaling pressure from the IWC has softened over time. One important reason is that the pro-whaling nations have used the same recruitment tactics as the anti-whaling nations: new pro-whaling nations have joined the IWC and some anti-whaling nations have left. Today, there is close to a 50/50 vote on some issues. This development indicates that majority rules can be a double-edged sword even for actors that are initially part of the majority, because participation or positions may change over time. In short, decision rules emerge as an important factor in determining receptivity. Consensus combined with 'fast-track' options has facilitated high goal attainment, whereas the advantageousness of a majority rule has depended on Norway's position with respect to the majority in the ozone regime and the IWC.

The *strength* of the core regime was seen as a function of legal status, specificity, level of ambition and compliance control. We have assumed that strong regimes increase the probability that the behaviour of foreign target groups will change in the 'right' direction according to regime objectives.[5] The IWC, a strong regime based on international law, had an immediate impact on Norwegian whaling policy when it started to tighten up regulations. This led to a sharp reduction in Norwegian quotas. The IWC thus shows that the extent to which regime strength promotes Norwegian goal attainment depends on Norway's position in relation to regime objectives. In the five other cases, Norway would benefit from a strong regime in terms of international goal attainment. Norway would clearly benefit if outsiders – i.e. other states and non-state target groups – changed those aspects of their behaviour that affect Norway negatively. Here it should be borne in mind that Norway may have other interests abroad than at home: for instance, a net importer of pollution would prefer that others contribute more towards reducing discharges, while the actor itself contributes less.

According to the criteria selected in this study, the climate regime and the CBD can be categorized as relatively weaker regimes than the ozone, air and marine pollution regimes. The commitment to reduce greenhouse gases under the UNFCCC is not binding, and the CBD is less specific and ambitious in terms of joint international commitments. Even though the climate regime has been strengthened by the Kyoto Protocol, the latter is not yet in force and has been severely weakened after the US withdrawal. This indicates that the high level of congruence between Norwegian positions and the climate regime has not, as could have been expected, produced a corresponding change in the behaviour of target groups outside Norway. Greenhouse gas emissions in the highly industrialized countries rose by eight per cent between 1990 and 2000, and are expected to increase by another eight per cent from 2000 to 2010 despite domestic measures currently in place to limit them (UNFCCC, 2003). The consequences of the CBD, on the other hand, are extremely difficult to assess. However, the CBD is equipped with an incentive mechanism through GEF, which has facilitated implementation particularly in developing countries. It is also strengthened by international follow-up activities, such as its Cartagena Protocol on biosafety and the synergetic effects of the FAO treaty on plant genetic resources for food and agriculture. To the extent that implementation actually leads to greater diversity compared to a hypothetical situation without the CBD, the results will benefit Norway indirectly as a consumer of pharmaceutical, agricultural and other products.

The ozone regime is strong in terms of legal status, specificity, level of ambition and compliance control. It has also been described as innovative, dynamic and flexible, particularly with regard to its Multilateral Fund aimed at supporting developing countries as well as the establishment of an implementation committee to assist the parties in cases of non-compliance. There is a clear link between regime strength and results: CFCs and halons were almost entirely phased out by 1995 – only eight years after the regime was established. Thus, the ozone regime has facilitated Norwegian goal attainment internationally by changing the behaviour of target groups located outside Norway. This has been particularly important for Norway since the depletion of the ozone layer is most acute in high latitudes.

The strength of the air pollution regime has also increased over time. More stringent and specific commitments backed up by strengthened verification and compliance mechanisms have been important for the reductions witnessed in transboundary air pollution (although domestic factors not directly related to the regime have also been central, not least in the SO_2 context). The convention has been particularly important for Norway by putting pressure on such major European emitters as the UK and Germany. The North Sea regime is comparable to the air pollution regime in terms of strength, and significant reductions in controlled substances have been achieved by most parties. As a net importer of marine pollution, Norway has benefited significantly from this development.

Our main conclusion here is that decision rules based on consensus and fast-track provisions on one hand, and high activity based on knowledge and political skill on the other hand, have facilitated a high level of international goal attainment. In most cases, Norwegian efforts have contributed to strengthening the core regimes. In turn, regime strength is related to change in the behaviour of foreign target groups, which has benefited the goals and ultimately the environmental quality of Norway.

Linked regimes: Congruence and venue shopping. High regime density is frequently seen as a problem that leads to duplication of work, coordination problems and contradictory obligations. From the perspective of one state, however, high regime density can create new opportunities. The main assumption is that Norwegian goal attainment internationally will increase when there is a high degree of congruence between the objectives of the core regime, the objectives of functionally linked regimes and Norway's positions. If there is a low degree of congruence between Norway's position and the objectives of the core regime, linked regimes may facilitate 'venue shopping', whereby states can choose institutional arenas and take advantage of existing regimes to promote their own agendas.

The IWC displays a clear mismatch between the objectives of the core regime and Norway's position. The whaling case serves as an example of 'venue shopping', since the IWC is linked to several other regimes which have increasingly supported Norway's whaling policy, partly against the objectives of the IWC (see Chapter 3). For example, GATT/WTO and UNCLOS III have strengthened the Norwegian position by recognising that whale management should fall under the jurisdiction of coastal states and by making sanctions against whaling nations more difficult. The 1973 Convention on Trade in Endangered Species of Wild Fauna and Flora (CITES) has also lined up with Norway's interests by supporting the Norwegian proposal to down-list the North-East Atlantic minke whale from the threatened species list.[6] Moreover, Norway contributed to the establishment of NAMMCO, which initially was seen as a management alternative to the IWC. The whaling case thus illustrates that whenever the interests of one specific state diverge significantly from the objectives of the core regime in a situation of high institutional density, that state can further its interests in linked regimes.

Likewise, the North Sea case illustrates the same phenomenon. In the 1980s, the EU Commission was the major 'blocker' of proposals for pollution abatement within the Paris Commission on land-based sources. One major reason was the lack of a firm legal foundation on environmental matters within the EU. This obstacle was gradually removed by the establishment of the North Sea conferences as an alternative institutional arena based on 'soft law' that resolved the legal problem for the EU. This move was undertaken by Germany, but it increased the influence and goal attainment of the 'pushers', including Norway.

In the case of ozone depletion, trade measures included in the Montreal Protocol on the ozone layer have not clashed with GATT/WTO rules, despite some initial fear of a potential conflict between trade liberalisation and ozone policy. Other international regimes have generally supported the Montreal Protocol and the position taken by Norway. The 1979 Convention on Long-Range Transboundary Air Pollution served as a learning experience for Norwegian and Nordic negotiators in the ozone negotiations. In the climate regime, the precedent set by CLRTAP made it easier for Norway to get acceptance for differentiated obligations. The second sulphur protocol, signed in 1994, served as a model for differentiated emissions commitments in the Kyoto Protocol. In the case of air pollution, other international agreements have neither conflicted with nor facilitated Norway's positions on international air policy.

The CBD could build on a number of other regimes within the nature conservation cluster, such as CITES and the Ramsar Convention on Wetlands (see Chapter 8). These conventions all pull in the same conservation direction, although their overlap did not give rise to turf-wars during the CBD negotiations. However, the parallel negotiations on intellectual property rights (TRIPs) under the GATT Uruguay Round caused some worries. TRIPs promotes the privatisation of genetic resources through intellectual property rights, while the CBD emphasizes equitable sharing and conservation. This complicated the negotiations, but Norway's position on TRIPs was in accord with the CBD negotiations. The EU patent directive has also been criticised for undermining the CBD objectives, and the Norwegian government's decision to accept the patent directive has been subject to major criticism, primarily on this ground.

Linked regimes have mostly had a positive impact on Norway's international goal attainment. They have mainly supported the objectives of the core regimes, served as learning experiences and models by diffusing regulatory ideas, or provided opportunities for 'venue shopping' in those cases where regime objectives have conflicted with Norway's positions.

In Table 9.2 (on the next page) we find a high level of support for the assumptions derived from the analytical framework (see Chapter 2). On the whole, the empirical patterns match our expectations fairly well.

The Downstream Process: Influencing Nationals

Goal attainment at home is conceived of as a function of three determinants: (1) the strength of core regimes and their congruence with domestic environmental policy goals; (2) the impact of other functionally linked regimes; and (3) the capacity of domestic institutions to integrate environmental policy goals into relevant 'cause' sectors, implement adequate policy instruments, and involve affected actors at an early stage in the decision-making process. Note that the whaling case is excluded downstream since goal attainment in this case is almost exclusively a matter of affecting actors outside Norway.

Table 9.2 Norwegian goal attainment internationally: expected vs. actual in relative terms[1]

Cases	Coherency in positions	Receptivity of core regime	Strength of core regime	Activity and leadership	Convergence between linked regimes	Expected goal attainment	Actual goal attainment
Marine	H	H[c]	Strong[c]	M/H	H[c]	H	H[c]
Air	H	H	Strong[c]	H	H	H	H
Ozone	H	M[a]	Strong	M/H	H	H	H
Biodiversity	H/M[c]	H	Intermediate	M/H	M/H	M/H	H
Climate	M/H	H	Intermediate	M	H	M/H	H
Whaling	M[c]	M[b]	Strong	H	M[c]	M	M[c]

H: High; M: Medium; L: Low.

[a] Majority voting enhanced Norway's goal attainment in this case

[b] Majority voting restricted Norway's goal attainment in this case.

[c] Significant changes in goal attainment have occurred over time.

[d] Significant differences in goal attainment between various components or substances.

Core regimes: *pressure on 'pusher'*. Demonstrating enthusiasm for joint action by pushing for strong regime commitments that affect foreign target groups does not necessarily mean that a state has high incentives to comply with its own commitments. Moreover, international regimes can pick up momentum that can 'backfire' on initially ambitious parties. We assumed that the impact of international regimes on domestic goal attainment in 'pusher' states would depend upon a combination of regime strength and the match between regime objectives and domestic environmental policy goals. If national goals are going in the same direction as regime objectives, but are less ambitious than international commitments, a strong regime will promote compliance with international obligations.

To increase the number of observations, we distinguish between various regime components in the following discussion. This modification provides us with three cases in the 'strong regime/high congruence' category: the ozone regime, the CLRTAP regime on SO_2, and the North Sea regime on dumping and incineration at sea (and hazardous substances to some extent). With regard to cases in this category, we assumed that regime influence would be limited: Norway would affect the regime, rather than the converse. This expectation roughly matches the empirical patterns observed. The ozone regime has previously been characterised as a particularly strong regime. The Montreal Protocol and later amendments have, however, had only a limited effect on Norway's ability to implement domestic measures to reach the set goals. The arrow of influence has mainly pointed the other way: Norway's efforts have influenced regime commitments. Still, the international ozone regime picked up momentum over time and affected goal attainment in Norway. Of particular importance is that the ozone regime has secured an equal competitive framework for affected industries in Norway.

The impact of the CLRTAP regime varies with different substances. In the case of SO_2 and the first Sulphur Protocol, Norway influenced the international commitments rather than the other way around. Most of the 30 per cent emission reduction was achieved by Norway *before* the protocol was signed. Domestic circumstances explain the high level of goal attainment on this specific substance. Exactly the same pattern is evident for dumping and incineration in the North Sea: Norway pushed hard for a ban on dumping internationally on the basis that dumping at sea has never been seen as a viable disposal option in Norway.[7]

The next, and perhaps most interesting, category is the combination of strong regimes and low congruence between regime objectives and domestic goals. In such cases, we expect that international regimes will affect Norway's goal attainment in the direction of compliance with international obligations (given that national goals are going in the same direction as regime objectives). Three cases fit the 'strong regime/low congruence' category: the CLRTAP regime on NO_x and on VOCs and the North Sea regime on nutrients. With NO_x and VOCs, the joint international CLRTAP commitments placed some pressure on Norwegian follow-up measures. In 1988, the Pollution Control Authority initiated work on a specific implementation plan to achieve the 30 per cent NO_x Declaration target. Likewise, various actions for reaching the 30 per cent reduction target were discussed for VOCs. Norway's environmental authorities have thus tried to use the international commitments to establish effective national policies, albeit with limited success in the 1990s. Likewise, the North Sea regime led to a detailed action plan on how to reach the joint commitments on nutrients.

The last category comprises cases characterised by 'weak regimes/high congruence'. We assumed that, in this situation, a regime would have limited influence on Norwegian goal attainment. The climate regime and the CBD fit this category. The climate regime has not placed any strong pressure on Norway to reduce GHG emissions. The stabilisation commitment included in the UNFCCC was not binding, and it was not taken seriously by Norway. However, the regime has influenced Norwegian climate policy by requiring national GHG emission inventories and national reports on action.

The CBD obliges its contracting parties to develop national strategies, plans and programmes for conservation and sustainable use, integrate conservation and sustainable use of biological diversity into relevant sector plans and policies, and develop systems of protected areas. The direct effect on Norway's forest policy is, however, limited. Although the CBD is a legally binding convention, the formulations 'as far as possible' and 'as appropriate' imply that there are hardly any obligations that need to be fulfilled. The CBD may have put extra pressure on Norway to protect its forest sector, which in turn has led to more protected areas – from 0.8 per cent to one per cent. In addition, the CBD has stimulated the integration of biological diversity in sector plans, with the Ministry of the Environment as coordinator.

The main deviation from our expectations is that most regimes have had a wider range of consequences than expected, which in turn implies that domestic target groups would have demonstrated a lower level of behavioural change in the absence of these core regimes. With the possible exception of CLRTAP and SO_2, all regimes have facilitated domestic goal attainment to differing degrees and in various ways. This observation also implies that international regimes tend to put pressure even on states that began as 'pushers' for stringent joint commitments.

In Chapter 2 we made a distinction between different pathways through which international regimes can affect domestic environmental policy. With the *legal* pathway, international rules and procedures become institutionalized into the domestic process by being incorporated into national law. The legal pathway does not seem particularly important to any of the core regimes, but it is crucial for understanding the influence of the EU (see next section). What does seem important is that many regimes have instigated domestic action programmes on how to reach the targets, and such action programmes normally contain a set of policy instruments linked to a time schedule. The *political* pathway has been in operation in various cases. First, the international regimes have empowered non-state actors in domestic policy debates. One example is the CBD, which has been repeatedly used by environmental organisations to put pressure on the government. Second, the Ministry of the Environment and its related agencies used joint international commitments to put pressure on 'reluctant' ministries and sectors. The North Sea Declarations are illustrative here: The Ministry of the Environment used the commitments on nutrients to put pressure to bear on the municipal and agricultural sectors in Norway.

Linked regimes: more pressure on 'pusher'. Linked regimes that facilitate goal attainment internationally will not necessarily facilitate goal attainment domestically. On the other hand, linked regimes that constrain goal attainment internationally may serve to further goal attainment domestically. New regimes may emerge and old may evolve, influencing domestic implementation. The 1994 European Economic Area (EEA) Agreement exemplifies how new institutions may emerge and affect implementation after initial positions have been developed in various core regimes. As a result of this agreement between Norway and the EU, Norway is obliged to implement most EU environmental legislation without having access to the main decision-making bodies in the EU system. In such cases, there is a high probability for conflict between linked institutional obligations and national environmental goals, and this may have a negative effect on domestic goal attainment.

With regard to ozone depletion, other related regimes have mainly facilitated domestic implementation. Two substances under the Montreal Protocol are also covered by the regime for protecting the North-East Atlantic (see Chapter 6). As both agreements aim at phasing out the same substances, but for different reasons, regime linkages have supported Norwegian goal attainment in both issue areas. Due to the EEA Agreement, Norway is also affected by the ozone policy of the EU. In general, EU regulations have forced a minority of its member states to agree on faster phase-out schedules. In 2000, the EU adopted a new regulation with stricter restrictions on import and consumption of ODS than those of Norway. This piece of legislation is likely to facilitate domestic implementation of stated Norwegian ozone goals. The ozone regime has also interacted with the climate regime, but mainly in the sense that the previously established ozone regime has influenced the climate regime (see below).

In the case of air pollution, the UNFCCC led to a coordinated domestic follow-up process on GHGs and NO_x emissions. While air pollution and global climate change represent different challenges, abatement efforts have been linked domestically, as CO_2 reduction measures generally led to a reduction of NO_x emissions as well. Norwegian climate commitments thus interact positively with NO_x commitments. The International Maritime Organisation (IMO) has also been involved with regard to ship-air emissions. In this case, the Ministry of the Environment made an effort to use the IMO to further domestic integration of environmental air pollution concerns in the Ministry of Transport. EU air pollution directives have had a positive impact on Norway in the case of VOCs and air quality. For example, an EU Directive led to a Norwegian regulation of VOC emissions from solvent-using industries (see Chapter 5). Likewise, the adoption of air-quality limits in 1997 was rooted in two specific EU directives. The strongest (positive) impact from EU air policy can be expected on the basis of the 1996 EU Air Quality Framework Directive and subsequent daughter directives, together with effects flowing from the 1996 Integrated Pollution Prevention Control (IPPC) Directive.

The CBD interacts with the climate regime, which has the potential to create both synergy and conflict with regard to Norwegian forest management. On one hand, there is a win-win situation between forest conservation and climate policy, since forests represent the most important 'sink' for CO_2 in Norway. On the other hand, the proposed measures to increase the 'sink' potential – including afforestation, reforestation and change of tree species – conflict with biodiversity concerns. Conflict has been avoided, however, because Norway has decided that domestic and foreign forestation projects should not be part of its national climate strategy and has called for a more formal cooperation between the CBD and the UNFCCC in order to avoid negative effects from forestry activities under the Kyoto Protocol. The EU Natura 2000 and the Habitat Directives are largely outside the EEA Agreement, but may indirectly push Norway towards establishing more protected areas. It is most likely that realisation of policy goals for protected areas would have been improved if Norway had had to abide directly with the Habitat Directive.

The climate regime interacts with several other international regimes, which are generally not an obstacle to Norway's climate policy goals. Measures taken to reduce emissions of air pollutants within CLRTAP affect energy efficiencies in different ways, but have overall had a positive effect on Norwegian goal attainment in the area of climate change. As stated earlier, the CBD could have a positive effect on Norwegian climate policy, since forest conservation increases the 'sink' potential. The ozone regime is also largely well coordinated with the climate regime in the sense that gases included in the Montreal Protocol are excluded from the climate regime. However, as noted, some substitutes for ozone-depleting substances (HCFCs, HFCs) are also powerful greenhouse gases and are thus regulated by the Kyoto Protocol. As Norwegian consumption of HFCs is estimated to grow considerably, the increased use of substitutes for ODS can lead to an increase in greenhouse gas emissions. However, Norwegian authorities expect that it will become technologically possible to develop new substitutes at a low cost in the near future. In practical terms, therefore, the potential negative interaction between the ozone and climate regimes is expected to be temporary, and is not perceived by decision-makers as a significant problem in Norway's domestic climate policy.

The climate regime is also linked vertically to the climate policy of the EU and the EU directive on emissions trading. This directive may create problems for Norway in two ways. First, the allocation mechanism for quotas differs between the EU (90 per cent free of charge) and the Norwegian system (sold or auctioned), so that those sectors in Norway currently paying the CO_2 tax would experience a significant reduction in abatement costs in a transition to an emissions trading system with an allocation of free permits. Second, the scopes of the two systems differ in a way that may make it difficult to include the Norwegian process industries in the Norwegian system from 2005 (see Chapter 7).

The EU has clearly been the most important institution linked to marine pollution. On the one hand, the North Sea Declarations have speeded up decision-making within the EU. On the other hand, EU directives have facilitated goal attainment for Norway with regard to nutrients and hazardous substances. The recent EU Water Framework Directive will have significant impact on Norway. First, Norway has to establish new administrative borders in order to coordinate water management within new river basin districts based on an ecosystem approach. Second, environmental targets have to be related to all water resources, including surface water, ground water and coastal water. The Water Framework Directive has been well coordinated with the North Sea regime in the sense that priority hazardous substances are based on those substances previously included in the Paris Convention and the North Sea Declarations.

Thus, other linked regimes and the EU have generally had a positive impact on Norwegian goal attainment domestically. Even cases that have received significant attention because of problems caused by linkages at international level – such as the links between the climate regime, the biodiversity regime and the ozone regime – appear less problematic from the perspective of one state. Norway has, for example, decided that domestic and foreign forestation projects should not be part of its national climate strategy, and ODS substitutes that are climate gases will be replaced by less harmful substitutes. Coordination to avoid negative consequences from regime linkages may in turn be traced back to the fact that one Norwegian ministry – the Ministry of the Environment with its subordinate agencies – has been in charge of most of these regimes.

Vertical interaction with EU directives has also generally had a positive impact on Norwegian goal attainment domestically, with the exception of a potential disruption identified at the interface between emission trading systems in the EU and Norway. This general finding is contrary to our assumption, which was based on the circumstance that Norway has limited influence on EU decision-making through the EEA Agreement. There are at least three possible explanations: First, Norway has followed a strategy of adapting to EU environmental legislation since the late 1980s, i.e. before the EEA agreement was in place (Dahl, 1999). Norwegian adaptation reflects the general economic and political importance of the EU for Norway. This strategy has led to few instances of significant divergence between EU directives and Norwegian objectives. Second, since the mid 1990s, EU legislation has tended to be more ambitious, or more substantive in terms of implementation, than Norwegian goals, legislation and other policy instruments. For example, it was an EU directive that led Norway to adopt a regulation concerning VOC emissions from solvent-using industries. Third, Norway may have had more influence on the shape and content of EU directives than expected against the backdrop of the EEA treaty. As the EU is a party to most international regimes that are important to Norway, this means that Norway can affect EU legislation through its participation in these international regimes. The North Sea cooperation provides us with a specific example of this mechanism. Since the 1980s, Norway and the Nordic countries have worked for a ban on dumping in the North-East Atlantic. This effort led to an international ban of this practice, adopted at the international North Sea Conference in 1987 and later within the framework of the OSPAR Convention. The EU, which at the time had no dumping directive, responded to this development by banning sewage sludge dumping in its Urban Waste Water Directive adopted in 1992. In this way, Norway was able to upload its domestic standards to the EU level through linked international regimes.

Domestic institutions: sector integration, policy instruments and involvement. The impact of core regimes and linked regimes cannot explain why Norway scores relatively low on domestic goal attainment. The impact of core regimes and linked regimes were generally found to have a positive impact on Norway's goal attainment. The implication is that the causes for the relatively low level (compared to the international level) of goal attainment are probably found at the domestic level in Norway.

Sector integration and policy instruments. In Chapter 2, we focused on sector integration and policy instruments, developing two main assumptions. First, we assumed that the more sectors affected by implementation (vertically and horizontally) in one issue area, the more difficult it would be to integrate relevant goals and measures in affected ministries to achieve stated goals. Second, we assumed that the adoption of adequate policy instruments would be more problematic when goal attainment was conditioned by behavioural change linked to diffuse sources, as opposed to point sources. Because different sources would require a broad-based portfolio of policy instruments, goal attainment is likely to be most difficult under such circumstances.

In the case of ozone depletion, implementation has not affected significant sector interests, and the need for integrating ozone policy in other governmental sectors has been limited (see Chapter 4). Moreover, the principal target groups – companies in the dry-cleaning, cooling, plastics and electronics businesses – joined in one interest organisation to make their voice heard in the implementation process. A broad portfolio of policy instruments was applied, including subsidies, information, regulation and a threat to implement a tax if the industry did not meet national targets. In addition, the industry saw new business opportunities: Norwegian companies could gain a competitive edge over other countries since Norway was one of the first to impose strict regulations.

Despite pressure from CLRTAP and other related international commitments, Norway has experienced significant implementation problems with regard to targets on NO_x and VOC emissions (see Chapter 5). In the case of SO_2, emissions stem mainly from industry and mining. These represent point sources, and industry has been regulated by emission permits since the mid 1970s. In addition, taxes on fuels have contributed to the reductions witnessed. Reduction of NO_x represented a greater challenge, since emissions stem from point sources and diffuse sources, not least road and ship traffic as well as the petroleum sector. Various packages of policy instruments and measures were worked out, but few measures were adopted because of the differing positions among the ministries involved: the Ministry of Finance resisted more governmental spending; the Ministry of Transport resisted measures to reduce traffic; the Ministry of Energy resisted measures that affected oil production; and the Directorate for Shipping resisted regulation of coastal traffic. The problems experienced in reducing NO_x emissions can thus mainly be related to lack of sector integration. The various ministries had different positions in the implementation phase.

The situation was the opposite in the case of VOC emissions. Here we would have expected high goal attainment, since behavioural change in one point-source sector (oil companies) would have been sufficient to reach the stated goals. Norway's difficulties in reducing VOC emissions relate mainly to inadequate policy instruments. The environmental authorities relied heavily on the possible establishment of a voluntary agreement with several oil companies. This agreement was to be based on best available technology (BAT) to recycle VOC evaporation at shuttle tankers loading crude oil. Such activities accounted for about 60 per cent of total (non-methane) VOC emissions in Norway. Installation of BAT on all twenty relevant ships would mean reductions in the order of 70 per cent from each ship. In the end, however, some of the major oil companies refused to support the deal, and the agreement collapsed.

Climate change policies affect most sectors of society, and the problem is caused by a multitude of different sources (see Chapter 7). While different ministries managed to agree on the country's climate position internationally, at home there has been heated conflict between the same ministries. Various ministries have disagreed on the unilateral stabilisation target and the scope of the CO_2 tax. Internal disagreement has provided target groups with an unpredictable regulatory framework, delaying domestic action. In addition, the CO_2 tax, which has been the principal climate policy instrument in Norway, has had only limited effect on CO_2 emissions. More recently, however, domestic resistance has faded in light of the Kyoto Protocol and the rise of emission trading as the main policy instrument. These events have removed the sources of internal disagreement – the unilateral target and the tax.

The forest sector was selected in the case of biodiversity (see Chapter 8). Most implementation problems experienced in the forest sector are rooted in diffuse sources, as well as in conflicts of interest between the Ministry of the Environment and the Ministry of Agriculture and between the central and local levels. The position of the Ministry of Agriculture is in turn linked to the considerable economic interests involved in forestry. The conflict between the two ministries has led decision-makers to adopt inadequate policy instruments. In fact, the economic instruments applied in the forestry sector, such as subsidies to build logging roads, may lead to loss of biodiversity.

Norway experienced significant problems in its efforts to reduce nitrogen emissions to the marine environment (see Chapter 6). Nitrogen emissions stem mainly from the agricultural and municipal sectors, necessitating horizontal as well as vertical integration of environmental concerns. Since 1992, the Ministry of Agriculture has placed significant emphasis on the national goals linked to Norway's North Sea commitments. A similar acceptance of the 'sector responsibility' principle has taken place at the county and local levels. Even though the agricultural authorities have increasingly integrated water and marine pollution concerns, there has been too little too late. Moreover, some measures to counter agricultural run off were implemented in the 'wrong' geographical areas, and the tax on commercial fertilizer was abandoned rather than increased.

Municipalities are permitted to finance investments in waste-water treatment through local taxes, and they are responsible for necessary purification. This responsibility was used to postpone investments in nitrogen removal. All affected municipalities submitted appeals to the Ministry of the Environment concerning the county governor offices' requirement that they be responsible for nitrogen removal. Eventually, fierce opposition at the local level contributed significantly to the cancelling of planned investments at 24 plants. In essence, vertical disintegration due to local resistance is one major cause of implementation problems in the case of nutrient emissions to the marine environment. With regard to hazardous substances, permits applied at point sources have proved more effective and led to a higher degree of goal attainment than was the case for nutrients.

Linking foreign and national policy: involvement of affected actors. In environmental policy, there is often a lengthy time-lag between Norway's actions abroad and at home. In 1988, Norway signed a political declaration on NO_x, which was to be achieved domestically ten years later. In 1987, Norway agreed to several North Sea targets – to be reached by 1995 or later. In 1997, Norway signed the Kyoto Protocol, to be implemented between 2008 and 2012.

In Chapter 2, we assumed that involvement of affected ministries and target groups at an early stage, ideally *before* national positions are developed and goals are fixed, would enhance their support in the implementation phase. We are thus interested in the extent to which the agencies and target groups in charge of implementation have been involved in the formulation of national goals and foreign policy positions at an early stage. We expect that early *access* for and *participation* by affected actors will facilitate goal attainment.

Access and participation add to our understanding of why domestic goal attainment varies across issue areas. In the case of ozone depletion, affected target groups were actually consulted all the way. The organization of target groups in one major interest organisation (KBF) facilitated consultation between government and industry. Notice that target-group involvement in the ozone case did not lead to a lower level of ambition in Norway's ozone policy, but target-group participation did affect how the results were achieved. The interest organisation was able to make the suggested tax conditional upon non-compliance with the set goals.

The CBD case shows that, at first, the Ministry of Agriculture did not consider the biodiversity negotiations to be particularly relevant. Later, however, they stepped up their efforts and influenced implementation – against the positions of the ministries of the environment and foreign affairs. While the agricultural department of the Ministry of Agriculture was involved from an early stage, the forest department became involved only later on. The position of the agricultural department was in line with that of the ministries of the environment and foreign affairs, while the forest department was reluctant to implement biodiversity commitments in the forestry sector.

Late involvement has not been a problem with regard to air pollution, but Norway's goal attainment in the North Sea case could have been different if the affected ministries and targets groups had participated at an earlier stage. In contrast to industry, the municipal and agricultural sectors were not involved as 'core insiders'. The agricultural sector responded constructively, but too late; some five years elapsed from the time the goals were developed until the agricultural sector became seriously involved in meeting the targets. Equally important was the fact that affected sectors were not prepared for the economic and practical consequences of implementing the joint commitments adopted in 1987. The decision-makers operated within a 'veil of uncertainty' that facilitated the adoption of joint commitments, but would make implementation more difficult at a later stage.

Norway's 1989 stabilisation target on CO_2 emissions stands out as a mystery in retrospect. Why was this unrealistic goal adopted in the first place? Part of the explanation may lie in the narrow participation in the process leading up to the decision. The Ministry of the Environment was the dominant ministry during the first formative years. The affected governmental sectors were not involved in the parliamentary process in 1989, nor were the target groups. If these interests had been involved, it is less likely that the target would have been adopted in the first place. When these actors did become involved in the early 1990s, they influenced the Norwegian policy on differential obligations and flexible, cost-effective mechanisms. By implication, Norway gradually shifted its focus from reductions at home to reductions abroad.

The upshot of these observations is that access and participation in goal formulation have consequences for domestic goal attainment at a later stage. Most cases studied here also support the assumption that including affected actors at an early stage will increase the likelihood of goal attainment. A dilemma for decision-makers is that such early involvement may reduce the level of ambition. However, this was not the case with ozone. When national goals are closely linked to international obligations, decision makers have probably much to gain by broad involvement of affected actors early in the decision-making process.

In general, there appears to be a close match between expectations and actual goal attainment (Table 9.3). The main conclusion is, first, that the influence of core regimes and linked regimes has in most cases facilitated goal attainment domestically. This observation implies that in the absence of regime and EU pressure, target groups located inside Norway would probably have caused more environmental problems. Second, the relatively low level of goal attainment domestically is caused by insufficient sector integration, inadequate policy instruments and insufficient involvement of affected actors at an early stage in the decision-making process. The number of sectors affected horizontally in different issue areas is one major factor. Many cases show that various ministries and agencies have resisted the principle of 'sector responsibility' in practice. Agreement on foreign policy positions is no guarantee for agreement on domestic implementation. The good news is that such resistance can be overcome: in the North Sea case, marine pollution concerns have increasingly been incorporated in the agricultural sector at central, regional and local levels. In two cases, vertical implementation failures have been reported. In these cases, local competence at municipal level has been actively used to dash national plans of action. Such local responsibility is part of the decentralist tradition in Norway (Nausdalslid, 1994). Other studies also indicate that that local resistance to national injunctions derived from international obligations may be part of a general trend (Harsheim and Hovik, 1996; Hovik and Harsheim, 1996).

Table 9.3 Norwegian goal attainment domestically: expected vs. actual in relative terms

Cases	Regime strength[a]	Impact of linked regimes	Level of sector integration and adequacy of policy instruments	Inclusion of affected actors at early stage	Expected goal attainment	Actual goal attainment
Marine: hazardous and dumping	H	Positive	H	H	H	M/H
Marine: nutrients	H	Positive	M	M	M	M
Air: SO_2	H	Positive	H	H	H	H
Air: NOx	H	Positive	L	H	M	L/M
Air: VOC	H	Positive	L	H	M	L
Ozone	H	Positive	H	H	H	H
Climate	M	Positive and negative	L[b]	L	L	L
Biodiversity/ forest	M	Positive and negative	L/M	M	M	M

H: High, M: Medium. L: Low.

[a] Strong regimes will promote goal attainment in all cases since none are significantly at odds with regime objectives. However, the relative impact of regimes on domestic goal attainment is conditioned by the level of congruence between regime objectives and national goals.

[b] Significant changes in goal attainment have occurred over time.

Most cases indicate that diffuse sources are more difficult to regulate than point sources. However, the government is by no means a 'victim' to the types of sources. Policy instruments are clearly important for understanding success and failure in domestic goal attainment. In the NO_x case, we saw that the means were available, but the political willingness to adopt them was missing. In the VOC case, the poor results achieved can be traced back to the priority given to 'wrong' policy instruments. The CBD case further illustrates how policy instruments in target sectors can have unintended negative consequences: subsidies applied in the forest sector actually worked against biodiversity goals. Finally, the ozone case shows how domestic goals can be achieved under benign conditions: a broad-based portfolio of policy instruments was adopted to attack a simple problem in terms of sectors and sources.

Alternative Explanation: Problem Type

The assumptions derived from the institutional approach have gained high empirical support. However, different explanatory approaches can explain the same patterns by emphasizing different causal mechanisms. In this section, we narrow in on a complementary approach based on problem type, as defined by cost-benefit distribution and configuration of actor interests. Some problems can be solved more effectively than others simply because they are politically easier to deal with. Likewise, Norway may attain its goals in some issue areas simply because the underlying problems are easy to deal with compared to other issue areas. Since target groups are located both abroad and at home in almost all cases, we have selected two indicators that distinguish between 'malign' and 'benign' problem types. First, success in attaining international goals depends on *related-actor interests*, that is, the configuration of interests and power of other states in the core regime. It is reasonable to assume that Norway's goal attainment in international regimes will decrease when the opposing forces increase. Conversely, Norway's goal attainment will increase when supporting forces make up a 'winning coalition'. In particular, we assume that the positions of the US and the EU are especially important for Norwegian goal attainment internationally. Second, success in domestic goal attainment may simply be a question of *cost-benefit distribution*, that is, the size and distribution of costs and benefits among target groups. Implementation will be extremely difficult if costs are high and concentrated in target groups and when benefits are modest and distributed to the society at large. Implementation will be simplified if costs are low and well distributed, and benefits are large and concentrated within specific target groups.

In the case of whaling, the target groups are almost exclusively located outside Norway, with the US a prominent opponent to the Norwegian position in the IWC. The EU is not a party to the IWC, but has generally been in line with the restrictive US position. Norway has thus faced a malign political problem, which would indicate low goal attainment internationally. The role of the US is important in explaining Norway's 1986 decision to stop commercial whaling. However, since the basic US position on whaling has remained at odds with that of Norway, this approach cannot explain why Norway later decided to resume whaling. The case is even more interesting if we consider the aspect of cost-benefit distribution: domestic economic costs have been almost negatively related to goal attainment in the sense that whaling has only minor economic importance in Norway, whereas the resumption of whaling and export of whale products might even prove politically and economically expensive in terms of loss of reputation, sanctions and boycotts. Thus, the whaling case is a mystery for anyone applying a narrow cost-benefit analysis to understand why Norway resumed whaling.

In sharp contrast to whaling, the high level of goal attainment in Norway's ozone policy can be partly understood against the backdrop of a benign political problem. At the international level, Norway was initially part of a 'winning coalition' including countries such as Canada and the US, as well as UNEP, whereas the EU was a 'laggard' in this phase. When the EU joined this coalition, the Norwegian position was in line with that of all major states that controlled the production of ozone-depleting substances. Within Norway, the phasing out of CFCs, halons and other ODS has been estimated to cost target groups only approximately NOK 65 million a year. By contrast, Norwegian implementation costs related to EU Regulation 2037/00/EC have been estimated at between NOK 290 and 470 million. The ozone challenge is thus far from a 'coordination problem' characterised by compatible interests between affected actors. Nevertheless, the high level of goal attainment witnessed in this case is the result of a benign problem and high institutional capacity.

The US (with Canada) is a party to the CLRTAP Convention, but has kept a low profile due to limited exchange of air pollutants between Europe and North America. Such major EU states as Germany, the UK and France altered their positions and became more in line with Norwegian interests, for various reasons. For example, Germany changed from a 'laggard' to a 'pusher' in the international negotiations after the 'Waldsterben' incident. This change in positions of important EU member states paved the way for a progressive EU policy on air pollution. In addition, significant reductions of regulated substances have taken place in these countries for reasons quite unrelated to environmental policies, such as the 'Wall Fall' effects in Germany. In Norway, several unrelated factors made SO_2 reductions easy compared to NO_x and VOCs. For example, the closure of the copper mines in Sulitjelma in the far north led to a 10-12 per cent decrease in SO_2 emissions from 1980 to 1993. Nevertheless, the differences in goal attainment between VOCs and NO_x cannot be ascribed to differences in the distribution of costs and benefits among target groups. In fact, VOCs stand out as a comparatively more benign political problem than NO_x, even though Norway's goal attainment on VOCs has been characterised as an outright failure.

In the case of climate change, the US determined the outcome of the UNFCCC in 1992, and has been the dominant actor in the international negotiations. The compatibility between the US and Norwegian positions ensured a high match between the Protocol and Norwegian positions. Since 2001 and the US withdrawal from the Kyoto Protocol, the EU has assumed the leading role in international climate regime. For Norway, domestic resistance and opposition to climate-change measures based on high and concentrated costs can help to explain why so little has been achieved in terms of reducing the expected domestic growth in GHG emissions. This is only part of the explanation, however, since significant domestic reductions could have been achieved at low cost. Moreover, the CO_2 tax was implemented against the interests, and the vigorous opposition, of the strongest economic sector in Norway – the petroleum industry.

The UK has been Norway's most powerful opponent in the marine pollution regime as well as in EU water and marine policy. The majority of parties, among them Norway, managed to influence the UK to accept the precautionary principle, which in turn paved the way for a significantly stronger regime on marine pollution, particularly with regard to hazardous substances. At home, however, the distribution of costs and benefits among target groups can only to a limited extent explain the relative positions and achievements of the sectors involved. For example, Norwegian industry had incentives to oppose the goals, but emerged as the most supportive sector, delivering reductions in accordance with obligations.

In the case of the CBD, the position taken by Norway, along with the other Nordic countries and much of the Third World, was at odds with powerful opponents, such as the US and the EU. The US resistance to equitable sharing was based on its considerable economic interests in the biotechnology sector. Nevertheless, the Norwegian/Nordic/Third World position influenced the outcome of the CBD negotiations. Domestically, goal attainment carries a higher price than perhaps initially expected. The forestry sector is characterised by concentrated costs, and benefits widely distributed throughout society. Forest owners thus tend to oppose any governmental conservation efforts that are not followed by financial compensation. Lack of technological solutions also helps make this problem politically malign.

An interesting pattern occurs at the international level across the cases. First, the US and the EU are parties to five of the six core regimes, but their role and importance with regard to Norwegian goal attainment varies significantly. The US position has facilitated Norwegian goal attainment internationally in the climate and ozone regimes; the US position has been at odds with Norway in the CBD and the IWC; and the position of the EU/Germany has facilitated Norwegian goal attainment in the CLRTAP. In the North Sea regime, Norway was part of a majority which was able to influence the position of the UK. This pattern indicates that US and EU participation and positions in the core regimes can vary significantly in importance for Norway's goal attainment internationally. A high level of congruence between Norway's position and the positions of the US and the EU is not imperative for high Norwegian goal attainment internationally. This also holds true when the US and the EU have been in agreement against the interests of Norway in the IWC.

The same conclusion can be drawn at the domestic level. Predicting goal attainment domestically on the basis of distribution of costs and benefits among target groups would have yielded only limited success. There is a mixed relationship between distribution of costs and benefits and domestic goal attainment. The ozone case shows the best fit according to the cost-benefit assumptions. This case is characterised by relatively low but concentrated costs and possible benefits. The CLRTAP case shows that SO_2 reductions can be understood in light of this perspective, and to some extent in NO_x, whereas VOCs cannot. Concentrated costs are important in explaining why so little has been achieved domestically in the case of climate change, but this is only part of the picture. In the case of marine pollution control, the positions and achievements of the industrial, agricultural and municipal sectors varied significantly, but not systematically in line with distribution of costs and benefits. However, concentrated costs and widespread benefits combined with lack of technological solutions can explain why conservation policy has met with significant opposition in the forest sector.

In conclusion, problem types in terms of distribution of costs and benefits at the domestic level and configuration of actor interests at the international level complement the institutional approach in explaining Norwegian goal attainment.

Conclusions

Studies of regime effectiveness usually explore how regimes engage countries in regime objectives. Effectiveness has been examined by the extent to which regime objectives have been met, and by the impacts of international rules on relevant behaviour of regulating agencies and target groups. In this book, we have chosen the opposite approach. Our perspective is how states engage international regimes in order to pursue their national goals within a given issue area. This approach may add some new insight to the study of regime effectiveness.

The first analytical implication is related to the role of, and attention paid to, domestic administrative and political institutions in the study of international environmental cooperation. In the study of regime effectiveness, increased attention has been paid to the role of domestic institutions in the downstream process of implementing international obligations. But the role of domestic institutions in the upstream process of shaping foreign policy positions in international regimes has been neglected. This book has shown not only that domestic institutions are important for shaping foreign environmental policy, but also that the link between foreign environmental policy and domestic implementation is crucial for goal attainment. Who is involved, and to what extent, in developing national positions in international negotiations is important for the prospects of implementing joint international commitments at a later stage (see below).

The second implication is related to the study of the effectiveness of single regimes. Until recently, the study of regime effectiveness was based on the assumption that international regimes existed in isolation from each other. Much academic attention has been directed at how international regimes can beat 'the law of the least ambitious actor' by various 'fast-track' options and through such behavioural mechanisms as norms and incentives. This study has shown that, under certain conditions, international regimes can affect even the most ambitious states. Strong regimes can 'backfire' and put pressure even on 'pusher' states to implement measures they would not have chosen unilaterally. Such regime pressure can emerge either from the dynamic development of the regime itself, which may lead to more ambitious joint commitments than national goals, or from divergence between international ambitions and domestic goals. For instance, net importers of pollution may prefer that others contribute more, while the actor itself contributes less.

The third point is related to regime linkages. The study of how interaction between international regimes affects effectiveness has recently gained increased attention. The literature on regime linkages has tended to focus on the international level and has traditionally emphasized problems caused by regime congestion and density. From the perspective of one state under crossfire from a high and growing number of international regimes, one would perhaps expect that the problems caused by regime linkages would come out even more clearly. However, we have seen that international regimes functionally linked with core regimes have mainly proven to be mutually reinforcing, at least in the case of Norway. This may well be traced back to the fact that the Ministry of the Environment has been in charge of most of the interrelated international negotiation processes.

Conflicts between international regimes tend to be alleviated at the national level. For example, the potential disruption between the climate and biodiversity regimes has been met by a decision that domestic forestation projects should not be part of the national climate strategy, even though Norway has supported the sink option in the Kyoto Protocol. Of particular interest is the observation that high regime density can be an advantage when there is a mismatch between national goals and the objectives of core regimes. 'Venue shopping' can allow actors to promote their agenda in functionally linked regimes – as when Norway used CITES to weaken the policy of the IWC by supporting the Norwegian proposal to down-list the North-East Atlantic minke whale from the threatened species list. We have also seen that Norway has contributed to establishing alternative institutional arenas, as in the case of NAMMCO, or utilised alternative institutional arenas established by others, as in the case of the North Sea conferences on dumping and incineration in the North Sea.

The study of regime linkages is still in the formative stage, and the conclusions above should be seen as preliminary. More research is needed if we are to fully understand how (effectively) states deal with the high number of international environmental treaties. Norway is party to some 70 international environmental agreements, of which only a few (albeit perhaps the most important) have been included in this study. Moreover, the results may look different if one systematically includes international institutions covering other issue areas, such as trade.

Conditions for Goal Attainment Internationally

In Chapter 2, we formulated various institutional conditions that were likely to promote or restrain goal attainment internationally (Table 2.1). Our analysis has shown that some conditions need to be specified and modified against the backdrop of the empirical observations.

First, when it comes to institutionalization of foreign environmental policy, we expected that a coherent position is important because internal division can be exploited by other states promoting opposing positions. The need for coordination is likely to increase, and the probability of a coherent position is likely to decrease, the more ministries and agencies that are involved in the 'upstream' process. Level of coherence is likely to be determined by, first, the number and types of branches involved, and second, the distribution of competence among those branches. If competence is shared between different ministries responsible for the same regime, coordination problems are likely to arise.

We found that in general, a high level of coherency in foreign policy positions directed towards the core regimes goes hand in hand with a high level of goal attainment internationally. The *types* of ministries involved in developing Norwegian positions appear more important for coherency than the number. Of particular interest is the observation that sector ministries representing fisheries and agriculture can play an important role in shaping environmental foreign policy even though other ministries are in charge of formulating national positions in international environmental cooperation. Internal disagreement between ministries can be overcome by means of active governmental leadership at top political level, as we saw in the whaling case. A high level of coherence has been found when there is only one ministry in charge, the Ministry of the Environment. Moreover, not only the same ministry, but the same core of key persons within the ministry, has been in charge of interrelated international negotiation processes – probably a vital factor in avoiding potential coordination problems caused by regime linkages.

Second, in the case of international core regimes, Norway's influence on joint commitments depends first on the level of activity and *means* by which Norway seeks to exercise its influence internationally, and second on the *receptivity* of the regime. The degree to which the regime can affect foreign target groups will depend on the *strength* of the regime. Adequate means of influence, high regime receptivity in terms of decision rules based on consensus, and strong regime commitments were assumed to increase the probability of goal attainment internationally.

The cases show that in malign situations characterised by strong opposing forces in the core regime, a high level of negotiation activity in the form of entrepreneurial and knowledge-based leadership emerges as a necessary condition for goal attainment. Norway has exercised its interests on the basis of political and scientific skills, rather than tactics and threats. Majority decision rules can prove to be a double-edged sword even for those parties that initially constitute the majority, since participation and positions may change over time and thereby change the ratio between the majority and minority within core regimes. Whenever the positions of one state vary between the majority and minority in different regimes, the influence of that state on regime commitments (regime receptivity) is generally maximized under conditions of consensus linked to 'fast-track' options. Regime strength raises the probability of affecting foreign target groups in line with national goals, if these goals are in line with the objectives of the core regime. When national goals and regime objectives diverge, regime strength can be negatively related to goal attainment internationally. For example, the IWC initially forced Norway to quit whaling, which was contrary to Norwegian goals.

Third, we expected that a high degree of congruence between regime objectives and/or good opportunities for 'venue shopping' would promote high goal attainment internationally. We found that in most cases, there has been a high degree of congruence between the objectives of the core regimes and linked regimes that has reinforced the Norwegian position. In several cases, lessons learned from linked regimes established earlier have been applied in order to affect new regimes. For example, Norway used the second sulphur protocol under the CLRTAP to get acceptance for the principle of differentiated emission commitments in the Kyoto Protocol. As noted, a low match between national goals and the objectives of the core regimes has been countered by means of 'venue shopping'.

Finally, we expected that a problem type characterised by compatible interests between Norway and influential actors in the core regime would lead to high goal attainment. As expected, problem types do matter for explaining Norwegian goal attainment. However, the interests, power and positions of other states participating in the core regimes can only to varying degrees explain Norway's goal attainment internationally. For example, the US is a party to five of the core regimes, but its interests and importance concerning Norwegian goal attainment vary widely from a crucial condition to a major obstacle.

Conditions for Goal Attainment Domestically

In Chapter 2, we also formulated various institutional conditions that were likely to promote or restrain Norway's domestic goal attainment (Table 2.2). First, we expected that strong core regimes and high congruence between regime objectives and national goals would promote high goal attainment. The cases show that Norway has generally pushed for strong regime commitments that affect non-Norwegian actors. By implication, there has generally been a high degree of congruence between regime objectives and Norwegian goals. However, we have found that, as to the impact of core regimes on domestic implementation, the regimes have had a wider range of influence than expected. Regimes have contributed to 'uploading' national standards to the international level and improved the competitive situation for Norwegian industries; they have instigated domestic action programmes and policy instruments, and empowered domestic non-state actors as well as environmental agencies to take action. In the absence of regime pressure, Norway's goal attainment domestically would probably have been lower.

Second, we assumed that a high degree of congruence between linked regime objectives and access granted to linked regimes would increase the likelihood of domestic goal attainment. We found that linked regimes in general and EU Directives in particular have mainly had a positive effect on domestic goal attainment for Norway. Contrary to our expectation, positive influence from linked regimes has taken place even when the same state has not participated fully in the decision-making process of the linked institution. The EU has in general promoted Norwegian goal attainment. This can be explained by more ambitious EU policy and because Norway has had more influence on EU decision-making than expected, since the EU is a party to most of the international regimes that are important to Norway. This indicates that interaction and diffusion of standards between international institutions can promote influence in the absence of formal participation in functionally linked regimes.

Third, with regard to domestic institutions, we expected that Norway would have high ability to attain stated goals in cases where one domestic sector was affected and sector integration was not needed, permits were possible to apply on point sources and those actors responsible for implementation were included at an early stage. The empirical observations from the cases show that low goal attainment domestically is caused mainly by lack of sector integration and inadequate policy instruments. Diffuse sources are more difficult to regulate than point sources, and problems affecting many sectors are more difficult to cope with than 'single-sector' problems. The government is, however, not a 'victim' of the types of sources and sectors affected. In some cases, more has been achieved on diffuse sources involving many sectors, than on point sources involving one sector. Moreover, local resistance to implementation has been a major obstacle to goal attainment domestically in cases where the affected municipalities have possessed significant competence in the activity subject to regulation.

Agreement between different ministries on foreign-policy positions is no guarantee for agreement in the implementation phase. But the inclusion of various ministries (and their associated agencies) and affected target groups at an early stage will increase the probability for goal attainment domestically. Early involvement tends to affect *how* goals are reached, rather than the ambitiousness of the goals themselves. This seems particularly true when national goals are linked to international obligations. Under such circumstances, domestic target groups and reluctant agencies have few opportunities to water down national goals, whereas governments have everything to gain from including the affected actors at an early stage.

Finally, we assumed that problem types characterised by concentrated benefits and widespread costs would lead to supportive target groups that would increase the prospects for goal attainment. The cases indicate that the interests and positions of domestic target groups derived from distribution of costs and benefits can only to some extent explain Norway's domestic goal attainment. If we had tried to predict domestic goal attainment on the basis of distribution of costs and benefits among target groups, our success would have been limited. The best example is probably Norwegian whaling policy, which is rooted in ideal rather than material interests.

This study has made an effort to add some new insight to the study of regime effectiveness by exploring the complex links between institutions at various levels of international environmental governance. From the perspective of one state with a reputation as a green 'pusher', we have seen that international regimes are not only capable of beating the 'the law of the least ambitious actor'. International regimes can affect and strengthen the environmental policy of even the most ambitious actors in international environmental cooperation. This book shows that a state with a coherent environmental policy will mainly benefit environmentally from the crossfire of a high number of interrelated international regimes.

Notes

1. Norway is allowed to let national emissions of greenhouse gases increase by one per cent by 2010 compared to corresponding emissions in 1990.
2. There has been a recent change towards more disagreement in the case of biodiversity.
3. For example, Per Bakken, Harald Dovland and Jan Thompson have been in charge of the air, climate and ozone negotiations in the 1990s.
4. Accordingly, this case is more in line with leadership approaches to foreign environmental policy than with institutional approaches (cf. Ch. 1).
5. Note that the purpose here was to conduct a preliminary empirical assessment of this relationship. A fully fledged empirical assessment of regime effectiveness, including causal relationships, would be beyond the scope of this study.
6. It should be noted, however, that if Norway should join the EU, this would make export of whale products extremely difficult.
7. However, Norway did deliver some waste for incineration at sea – a practice that was terminated in 1989.

References

Dahl, A. (1999), 'Miljøpolitikk – full tilpasning uten politisk debatt', in D.H. Claes and B.S. Tranøy (eds), *Utenfor, annerledes og suveren? Norge under EØS-avtalen*, Fagbokforlaget, Bergen.

Harsheim, J. and Hovik, S. (1996), *From Global to Local Perspective: A Study of Local Implementation of the North Sea Declarations*, NIBR Report No. 6, Norsk Institutt for By og Regionsforkning, Oslo.

Hovik, S. and Harsheim, J. (1996), *Environmental Policy in Local Politics*, NIBR Report No. 1996:5, Norsk Institutt for By og Regionsforkning, Oslo.

Naustdalslid, J. (1994), 'Innleiing: Globale miljøproblem-lokale løysingar', in J. Naustdalslid and S. Hovik (eds), *Lokalt miljøvern*, TANO/Norsk Institutt for By og Regionsforkning, Oslo.

OECD (1994), *Environmental Performance Reviews: Norway*, OECD, Paris.

OECD (2001), *Environmental Performance Reviews: Norway*, OECD, Paris.

UNFCCC (2003), 'Rich Countries See Higher Greenhouse Gas Emissions', press release, The UNFCCC Secretariat, Bonn.

Notes

1 [illegible]
2 [illegible]
3 [illegible]
4 [illegible]
5 [illegible]

References

Dahl, A. [illegible]

Hammond, [illegible]

Haq, [illegible]

Maclaren, [illegible]

[illegible]

OECD [illegible]

OECD [illegible]

[illegible]

Index